A SPECTRUM BOOK

PRENTICE-HALL, INC., Englewood Cliffs, New Jersey 07632

Don R. Dickson

Weather And Flight

AN INTRODUCTION TO METEOROLOGY FOR PILOTS

Library of Congress Cataloging in Publication Data

Dickson, Don R.
   Weather and flight.

   Bibliography: p.
   Includes index.
   1. Meteorology in aeronautics. I. Title.
TL556.D5   629.132'4   81-15772
                                AACR2

ISBN 0-13-947119-7

ISBN 0-13-947101-4 {PBK.}

        This Spectrum Book is available to businesses and organizations
        at a special discount when ordered in large quantities. For
        information, contact Prentice-Hall, Inc., General Publishing Division,
        Special Sales, Englewood Cliffs, N.J. 07632.

To my lovely wife, Myra, and in memory of a brave pilot, James E. Lemon, who sacrificed his life so his friends might live.

© 1982 by Prentice-Hall, Inc., Englewood Cliffs, New Jersey 07632

A SPECTRUM BOOK

All rights reserved. No part of this book
may be reproduced in any form or by any means
without permission in writing from the publisher.

10 9 8 7 6 5 4 3 2 1

Printed in the United States of America

Editorial/production supervision and interior design by Frank Moorman
Cover design by Jeannette Jacobs
Manufacturing buyer: Cathie Lenard

PRENTICE-HALL INTERNATIONAL, INC., *London*
PRENTICE-HALL OF AUSTRALIA PTY. LIMITED, *Sydney*
PRENTICE-HALL OF CANADA, LTD., *Toronto*
PRENTICE-HALL OF INDIA PRIVATE LIMITED, *New Delhi*
PRENTICE-HALL OF JAPAN, INC., *Tokyo*
PRENTICE-HALL OF SOUTHEAST ASIA PTE. LTD., *Singapore*
WHITEHALL BOOKS LIMITED, *Wellington, New Zealand*

CONTENTS

Preface, v

1. Introduction, 1
2. Atmospheric Heat, 14
3. Atmospheric Pressure and Its Influence, 26
4. Moisture, Clouds, Precipitation, and Other Hazards, 43
5. Air Masses, Fronts, Lows, and Highs, 57
6. Thunderstorms, 74
7. Turbulence, 91
8. Icing and IFR Weather, 100
9. Weather and Terrain Flying, 114

10 Weather to Soar, 123

11 Aviation Weather Assistance, 138

Bibliography, 149

Appendix: Glossary
of Weather Terms, 151

Index, 183

PREFACE

The study of aviation and the study of weather are inseparable. All pilots quickly learn how true this is and obtain at least an operational definition of the terms "aviation" and "weather." Such definitions are usually based on personal experience, so to avoid confusion, let us formally define these terms:

aviation: the art or science of flying airplanes; making and operating craft heavier than air.

weather: the general conditions of the atmosphere at a particular time and place. (More specifically, we shall define weather as the state of the atmosphere or, more precisely, the changing atmospheric conditions, especially as they affect man and his activities.)

Flying is one of man's activities that takes place entirely in the atmosphere, and aviation weather cannot be treated in a purely academic context because each aspect of weather relates to flight safety and aircraft operations. The purpose of this book is to help pilots gain a better understanding of weather in order to fly with safety and enjoyment. The ultimate goal of this book is to help reduce aviation-related accident and death statistics; pilots should die of old age.

One of the first things you will learn is that most weather is good. It is important to appreciate good weather, and also to recognize and respect marginal and hazardous flying conditions and avoid violent weather as you would the plague. You should always keep in mind that the atmosphere will be your friend if you understand it and treat it with respect. But it will be a deadly enemy for you and your flying machine if you don't. It is necessary for your safety and the safety of those who fly with you to recognize potentially marginal and dangerous weather situations. Early recognition of such situations gives you the opportunity to decide your plan of action before you are encumbered by, and emotionally involved in, hazardous weather conditions. This, in turn, will greatly improve your chances of a safe flight. A good rule of thumb to follow is "Never make any news that you will not be around to read." If you understand and use the weather, you will probably be a safe and happy pilot; if you disregard it and permit it to use you, make sure your will is written before you take off!

I am indebted to my wife Myra for unfailing patience and support, to George Batchelder of the Salt Lake City GADO office and the Salt Lake City Flight Service Station for suggestions and encouragement, to Madge Ross and Judy Hardy for their assistance, to Deanna Plumhof for typing the manuscript, and to Charles Quilter for his superb art work. Finally, I wish to thank Mary E. Kennan and the staff of Spectrum Books of Prentice-Hall for their unfailing cooperation.

D.R.D.

Professor Don R. Dickson teaches in the Meteorology Department at the University of Utah. He has written other books and an instructional television series in the field of meteorology for pilots and has published numerous technical reports.

1 INTRODUCTION

On a cool December morning in 1903, aviation meteorology was born. The Wright brothers had made the first successful powered flight in a heavier-than-air flying machine. Although the flight was epic-making and unprecedented, it wasn't the entire story. Recognizing that future flight would be impossible without careful monitoring of weather conditions, the brothers asked for the aid of weather bureau forecasters, and they got it. From that day on, precise reporting conditions became the rule at Kitty Hawk, North Carolina—and elsewhere—helping people to achieve their dreams of manned flight.

Why is the weather so vital to aviation? Think for a moment: Imagine you are floating in a rowboat in the middle of a vast ocean. A slight shift in the wind direction, bolstered by a gust of unexpected speed, sends a wave over your rowboat, nearly capsizing it for just a moment. Frightening? Certainly, and yet the atmosphere is nothing more or less than an ocean of air. A rowboat floating in the middle of an ocean is much like a small plane flying in the atmosphere. So can a responsible pilot afford to disregard the changing motions in a sea of air? The necessity to know, in this case about the atmosphere, is always a convenient place to begin. Isn't it preferable that the sea of air be less mysterious, yet more predictable?

Key words concerning the atmosphere are composition and

behavior; one affects the other. What makes up the air relates directly to the chemical and physical laws that govern its behavior and, incidentally, to the role that it plays in your flight.

COMPOSITION The earth has an oxidizing atmosphere that allows it to sustain life. The atmosphere is primarily composed of two gases: nitrogen (78%) and oxygen (21%). Certain minor elements, notably argon and carbon dioxide, make up the remaining 1%. Krypton (of "Superman" fame) is a trace element of lesser magnitude (see Table 1-1). When carbon dioxide and ozone are blended with water vapor, a significant contribution is made to the earth's heat budget. As these gases absorb prodigious amounts of long-wave radiation, they also heat the atmosphere.

Moisture exists in the atmosphere in three phases: solid (ice), liquid (water), and gaseous (water vapor). The actual amount of water vapor varies: Global concentration of gaseous moisture is greatest (4%) in humid equatorial climates; conversely, the poles and deserts contain very little. This variation in atmospheric water vapor is responsible for our weather.

TABLE 1-1 Composition of the Earth's Atmosphere

Constituent	Percent by Volume or by Numbers of Molecules of Dry Air
Nitrogen (N_2)	78.084%
Oxygen (O_2)	20.946%
Argon (A)	0.934%
Carbon dioxide (CO_2)	0.031%
Neon (Ne)	1.82×10^{-3}
Helium (He)	5.34×10^{-4}
Methane (CH_4)	1.5×10^{-4}
Krypton (Kr)	1.14×10^{-4}
Hydrogen (H_2)	5×10^{-5}
Nitrous Oxide (NO_2)	3×10^{-5}
Xenon (Xe)	8.7×10^{-6}
Carbon monoxide (CO)	10^{-5}
Ozone (O_3)	up to 10^{-5}
Water (average)	up to 1

Richard M. Goody, James C. G. Walker, *Atmospheres*, © 1972, p. 3. Reprinted by permission of Prentice-Hall, Inc., Englewood Cliffs, New Jersey.

GAS LAWS Well-mixed "air" can be treated as a perfect gas — obedient to the ideal gas laws. In England, Robert Boyle discovered that when a given amount of gas is kept at constant temperature, the product of the pressure and the volume remain constant. A century later French physicists, Jacques Charles and Joseph Gay-Lussac, recognized that nearly all matter expands when heated. In separate experiments they determined the volume expansion of a gas with increasing temperature while maintaining a constant pressure. This suggests that all molecular activity ceases, and that gas is unable to exert any pressure when the temperature reaches absolute zero (-273°C). If the individual laws of gas behavior formulated by Boyle, Charles, and Gay-Lussac are combined, a single gas law defining the relationship between pressure and temperature for a gas of any molecular weight is the result:

$$P = \rho R_d T \qquad (1\text{-}1)$$

where P is the pressure of the air, ρ is the density of the air, R_d is the gas constant for dry air, and T is the absolute temperature of the air.

Humidity, the amount of moisture in the atmosphere, plays an important role in determining the atmospheric density. Water vapor, with a molecular weight of 18, is lighter than air, with a molecular weight of 29. This means that the more water vapor contained in the air, the lighter that air will be. Moist air is more buoyant than dry air. This concept will be further explored in the section on atmospheric stability in Chapter 2.

VERTICAL STRUCTURE The atmosphere is divided into several layers (Figure 1-1). In a general sense, two large subdivisions exist: the homosphere is located from the earth's surface up to about 80-100 km; in this atmospheric region no gross changes occur in gas composition. The heterosphere is above the homosphere, and, unlike its counterpart, is characterized by a variation in the composition and mean molecular weight of its constituent gases.

A second system of describing the atmospheric layers makes use of the vertical temperature structure. These divisions, beginning at the earth's surface, are:

troposphere That region of the atmosphere in which the temperature decreases at an average lapse rate of 6°C per kilometer. The troposphere extends from the surface to 16 km at the equator, to 11 km at 50° latitude, and to 9 km at the poles.

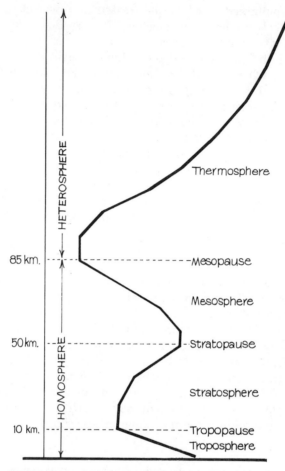

FIGURE 1-1 Major regions of the atmosphere and the ICAO standard atmosphere.

stratosphere Located upward from the troposphere, the temperature of the stratosphere is stable and warms to +7°C at 50 km.

mesosphere The mesosphere has a temperature lapse rate very similar to that of the troposphere, and similar temperature magnitudes are found, up to a ceiling of about 80 km.

thermosphere That layer of the atmosphere extending from the top of the mesosphere to the atmosphere-space boundary, in which the temperature increases as the height increases.

The regions of greatest interest to pilots, of course, are the troposphere and lower stratosphere. Most man-inspired flight is accomplished in the troposphere, and this atmospheric layer is where weather is found. The troposphere (from the Greek *tropo*, to turn) is the churning sphere that contains the major amount of the atmospheric mass. One-half of the atmosphere lies below 5.5 km (18,000 ft). The lowest stratum of the troposphere is the planetary boundary layer. In this layer all air motion shifts from the disturbed flow near the surface to the smooth frictionless flow of the free atmosphere. Frictional drag caused by surface roughness deflects the wind toward lower pressure.

Also, it is here that intense heating of the ground (by insolation) occurs. The air layer adjacent to the ground is also warmed by heat transfer from the ground source. The ground, in turn, does not warm uniformly because of its composition, configuration, and cover. This unequal heating causes what we might picture as air bubbles to form, allowing weather processes to be set in motion and heat to be transported from the warm surface to the air above. Thus, the air in this dynamic layer may be very unstable during the daytime. However, during the night the physical heating process (insolation) is no longer present, and terrestrial long-wave radiation from the earth's surface causes the layer above the surface to cool rather rapidly (see Figures 1-2 and 1-3*).

Pressure and density also have vertical structure, but these differ from the vertical structure of temperature. Pressure

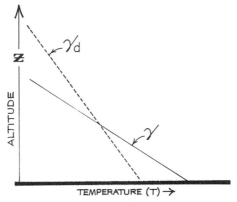

FIGURE 1-2 A super-adiabatic lapse rate.

---

*The Greek letters γ and γ_d in Figures 2 and 3 symbolize the ambient and dry adiabatic lapse rates, respectively. The term *lapse rate* refers to the change of temperature with height, and an *adiabatic* process is one in which heat does not enter or leave the system (the system, of course, in this case is the atmosphere). A more complete discussion is given in Chapter 2.

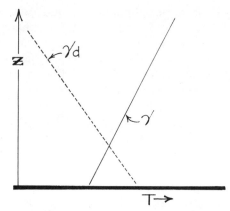
FIGURE 1-3 An inversion lapse rate.

and density decrease as height increases throughout the atmosphere, whereas temperature increases in the stratosphere and thermosphere.

That air has weight was demonstrated by Torricelli in 1643. At an elevation of 3,050 m (10,000 ft) above sea level, the atmosphere exerts only about 70 percent of the pressure that it does at sea level (see Table 1-2). This means that the oxygen available for breathing at that altitude is much rarer than closer to earth, at sea level, for instance. The rate of oxygen absorption is altered drastically, making it harder to take in sufficient quantities of the gas for survival. This is very important to a pilot, or indeed, to any-

TABLE 1-2 The Relationship between Pressure and Altitude in the Standard Atmosphere

Pressure Fraction of an Atmosphere	Altitude	
	Meters	Feet
1.00	0	0
0.83	1,525	5,000
0.69	3,050	10,000
0.63	3,660	12,000
0.56	4,570	15,000
0.50	5,490	18,000
0.46	6,096	20,000
0.37	7,620	25,000
0.30	9,144	30,000
0.24	10,668	35,000
0.18	12,192	40,000

one inside a plane at such heights. The effect of less air pressure can become physically evident in other ways. Distinct physiologic effects are apparent when the oxygen content of the blood is reduced, thereby lowering the brain's oxygen supply. A person usually becomes drowsy, feelings of exhaustion develop, judgment and vision gradually fade, and if the oxygen reduction continues unabated, fainting results. This gamut of symptoms is referred to as *hypoxia* and, like carbon monoxide poisoning, is extremely difficult to detect while it's happening.

So, if you plan to fly at or above 3,050 m (10,000 ft) for any prolonged period, it is wise to have supplemental oxygen aboard to guard against such dangers. If your aircraft is capable of flying above 7,600 m (25,000 ft) you will probably also want a pressurized cabin. At such heights, atmospheric pressure is not of sufficient magnitude to support life even if one is breathing pure oxygen.

It should be stated from the outset that the lack of oxygen in the air also reduces the efficiency of the airplane's engine, unless it is a supercharged one.

WEATHER ELEMENTS

Weather is affected to a significant degree by varying amounts of water vapor in the atmosphere. Without including water vapor, weather would consist of little more than wind and flying dust. In an atmosphere with varying amounts of water vapor, however, all sorts of weather phenomena can — and do — occur. Cloudiness, thunderstorms, turbulence, fog, rain, snow, freezing rain, and such combine to create a veritable Pandora's box of weather conditions, all posing unique hazards to aviation.

Facets of weather may be divided into two basic categories:

1. Conditions that can be detected by human senses: temperature, humidity, wind, clouds, precipitation in its various forms, thunder, and some atmospheric pollutants.
2. Conditions that cannot be detected by human senses: pressure and some atmospheric pollutants.

For many years, man's knowledge of weather was descriptive and limited to the components that could be sensed. The first observations of a weather element made by man involved precipitation and were dated as early as 2000 B.C. These records were transcribed in Sanskrit and uncovered only recently from the ancient Vedas, the holy books of old India. The first real quantitative measurements of weather, however, were made in the seventeenth century, and occurred with the invention of instruments capable of measuring the sensed elements of weather. Specifically, the thermometer could measure temperature just as the barometer and anemometer could define pressure and wind. When Morse invented the telegraph in the middle of the nineteenth century, meteorology really became viable. Now that a means of rapid communications had been established, a network of weather stations could be developed. Actually, the first meteorological observations in the United States were made by the U.S. Army Medical Corps during the Civil War. This was because physicians were among the better trained scientists of the times, and they had discovered that healing occurred faster when pleasant weather conditions were observed. From 1863 until the conclusion of the war, weather reporting greatly aided the Yankee cause, although the received information remained quite sketchy.

WIND Wind has both direction and speed. Each can play an important role in aircraft operations. Winds aloft were first observed by man, perhaps eons ago, as he watched the motions of clouds floating through the heavens. But other aspects of wind weren't so simple to observe. Wind direction was one of the first elements measured by the Greek philosophers in ancient Athens in about 350 B.C. A tower of the winds was built with eight vertical sides, each side facing one of the cardinal directions of the compass. Eight engraved figures decorated the sides, the figures representing winds from each direction. Perched astride the tower was a conch shell wind vane that must have looked impressive. Until Robert Hook, a British instrument maker, reintroduced the flat plate anemometer in 1664, wind speeds were not measured. There were great difficulties in perceiving weather elements at altitudes above the valleys and flatlands where most men lived, at least prior to the eighteenth century. Then, in 1753, the balloon was invented. The balloon provided a platform basic to mounting meteorological sensors, and these could be used to measure weather elements aloft. Prior to the invention of the balloon, such measurement of weather elements could only be done by transporting measuring instruments to the slopes or tops of mountains. There were many early examples of this tedious approach. One such case was that of Blaise Pascal; he discovered that barometric readings decreased as he ascended the 52-m Tour St. Jacques in Paris. Pascal's findings were later confirmed by Perier, who used a barometer to measure pressures at altitudes of almost 1,460 m. Scientific observations were made on the slopes of the Andes mountains, in order to determine the height of the freezing point at latitudes. The instruments used in these various expeditions could only be carried by the experimenter himself. Thus, it was recognized that in order to obtain measurements of the free atmosphere aloft, a carrier such as a kite or balloon must be used to lift the meteorological sensors skyward. At last, both man and his sensors had arrived in the heavens. Demonstrating this triumph in 1786, Jeffries and Blanchard (a professional aeronaut) carried a thermometer, a barometer, a hygrometer, a watch, and a compass aloft over London in a balloon. On this flight the pioneers made frequent random observations and inaugurated balloon use for weather observations. Balloon excursions into the higher atmosphere were soon made. In 1804, Gay-Lussac risked his life by rising to an estimated 6,636 m to obtain air samples. His subsequent observations of the vertical temperature gradient were accepted as being representative for years. Many balloon flights were made during the nineteenth century, but there were problems. Manned balloon flights were quite expensive, and unfortunately were limited to fair weather. The scientific community wanted to learn more about the air aloft during inclement weather, and began to consider kites for atmospheric studies. Franklin had made a kite famous in his celebrated experiment demonstrating the electrical nature of lightning. Alas, even kites weren't perfect. They had a limited range simply because the strings holding them would break. Thus, no systematic use of kites was made until 1885, when Archibald introduced steel piano wire instead of string as a means of reaching greater heights. A few years later, in 1893, Hargrave invented the box kite, which could lift a heavier instrument package into the heavens. After 1894, kites were used on a regular basis at the Blue Hill Observatory in Massachusetts. C. F. Marvin designed a kite meteograph, and 16 kite observation stations were eventually opened in the United States. The box kites transported temperature, humidity, and pressure sensors aloft. The upper winds were measured with hydrogen-filled rubber balloons and tracked by a theodolite measuring both direction and speed. The balloon observations were the best available but, alas, were insufficient. It soon became apparent that winds aloft could only be measured on relatively cloud-free days or perhaps below the cloud bases on overcast days. Most significant meteorological events did not occur on days when the sky was clear, but these observations provided much more information than could be obtained by simply watching clouds. Eventually, meteorological sensors attached to aircraft wings were used to measure weather conditions in the upper atmosphere. While these aerometeorolograms could not be used during bad weather, they were able to measure temperature and humidity and were an improvement over balloons and kites.

The radiosonde became the next important instrument used to sample atmospheric conditions. This sensor was first suggested by Hermite in 1892, but it encountered immediate problems, not the least of which was that the radio had not been invented yet! How could any one retrieve the balloon along with its valuable meteorological information? Even after the radio was invented, it took over 30 years to develop the radiosonde as a transmitter of meteorological data. Today, a sensor package on a balloon platform measures temperature, pressure, humidity, and winds aloft when used with a theodolite. These "rawinsondes" were a great advancement in the measurement of meteorological variables of the atmosphere and are applicable in all kinds of weather situations.

CLOUDS AND VISIBILITY

Knowing the various types of clouds and their distribution is useful for flight safety. The best information on clouds (other than direct pilot observation) will be at terminals, where extensive cloud observations and forecasts are made. Cloud observations are very important to flight operations, especially with regard to heights of bases and tops. A ceiling is the height measured from the surface to the base of the lowest opaque broken or overcast layer of clouds aloft; it is not measured for a cloud layer classified as thin. Occasionally it is reported as the vertical visibility in a surface-based obscuring phenomena. When ceilings cannot be measured with a ceiling light or ceilometer, or based on objects of known height in contact with the ceiling layer, they are reported as estimated.

Visibility is paramount for the pilot. It differentiates between whether you can fly or not, or whether flying is restricted to instrument flight rules (IFR). If you have an IFR rating and it is not current, you should not fly in "bad" weather. What is visibility? It is the greatest distance at which selected objects can be seen and identified. The visibility reported in the Service A weather reports is prevailing visibility, that is, "the greatest visibility equaled or exceeded throughout half the horizontal circle which need not necessarily be continuous, and runway visual range is the maximum distance in the direction of takeoff or landing at which specified light markers can be seen by a pilot at touchdown."*

Certain atmospheric phenomena can create poor visibilities. The most common IFR weather producers are fog, low clouds, and blowing phenomena such as dust, precipitation, or air pollution. Each of these meteorological situations cause specific problems to aviation, especially during landing and take-off. These meteorological phenomena will be discussed in detail in Chapter 4.

Various weather elements are observed at hundreds of weather stations in the United States. Many more are sighted in the world weather watch system of the World Meteorological Organization. These observations are then collected through a series of communication networks. At the National Meteorological Center (NMC) in the United States, all the information is fed into computers that analyze data, plot weather charts, and make numerical forecasts. The charts and forecasts are sent to various users: aviation and public forecast offices of the National Weather Service (NWS), Flight Service Stations (FSS), and other offices of the Federal Aviation Administration (FAA) which need and use the data. The NWS forecast centers and FAA Flight Service Stations are located throughout the United States. The aviation forecasts, prepared by the NWS forecast centers, are disseminated to the FSS and Air Route Traffic Control Centers (ARTCC). Eventually, this weather information is then made available to you, the pilot.

At NWS forecast centers, meteorological data are received from NMC as electronically produced weather charts that provide both current and prognostic weather information.

Clouds and cloud observations play an important role in aircraft safety. Aviators should learn to recognize clouds as signposts in the skies and have an understanding of their classifications. Clouds are observed from the ground as well as from aircraft and satellites. These cloud observations are plotted with symbols (Table 1-3), as indicated in the sample model shown in Figure 1-4.

The World Meteorological Organization recognizes four families of clouds: high, middle, low, and clouds of vertical development. Heights of the various clouds vary according to latitude. The low-level clouds are most regular for all latitudes,

---

*Departments of Commerce, Defense, and Transportation. *Federal Meteorological Handbook*, No. 1. "Surface Observations," Chapter A6-3, 2.2. Washington, D.C.: U.S. Government Printing Office, 1979.

TABLE 1-3 Weather Station Plotting Symbols

Symbol	Meaning
a	Characteristic of pressure tendency
C_H	Clouds of genera Ci, Cc, Cs
C_M	Clouds of genera Ac, As, Ns
C_L	Clouds of genera Sc, St, Cu, Cb
dd	True direction, in tens of degrees, from which wind is blowing
ff	Wind speed, in knots
N	Fraction of the celestial dome covered by cloud
PPP	Atmospheric pressure reduced to sea level (millibars)
pp	Amount of pressure tendency during past 3 months
RR	Amount of precipitation for the past 6 hours
TT	Temperature of air
$T_d T_d$	Temperature of dew point
ww	Present weather
$S_p S_p$	Special phenomena, general conditions
$s_p s_p$	Special phenomena, detailed description

with bases ranging from the earth's surface to 2 km (6,500 ft). The middle clouds have bases extending from 2 to 4 km (6,500 to 13,000 ft) in the temperate latitudes and from 2 to 8 km (6,500 to 25,000 ft) in the tropics. The bases of the high clouds stretch from 3 to 8 km (10,000 to 25,000 ft) in the high latitudes, from 5 to 13 km (16,500 to 45,000 ft) in the temperate latitudes, and from 6 to 18 km (20,000 to 60,000 ft) in the tropics.

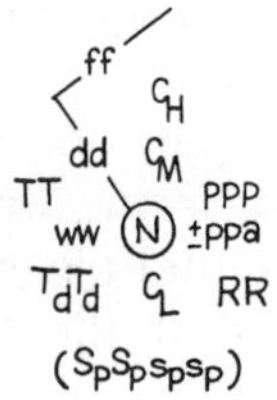

FIGURE 1-4 Plotted station model.

Clouds were first defined and classified by British scientist Luke Howard in 1803. Each cloud variant may be recognized as one of three basic types: cirrus, cumulus, and stratus — and these names have meaningful Latin derivations. *Cirrus* means a lock or tuff of hair; *cumulus* means a heap or pile; and *stratus* means a cover or blanket. Clouds that produce precipitation are called *nimbus* (meaning rainstorm or cloud), a name that is not recognized in the International Cloud Classification but is incorporated into the names of rain-producing clouds, for instance, cumulonimbus and nimbostratus.

The *International Cloud Atlas* of the World Meteorological Organization defines ten basic classifications of clouds, which are given according to the heights of their bases:

1. High clouds
    (a.) cirrus (Ci)
    (b.) cirrocumulus (Cc)
    (c.) cirrostratus (Cs)
2. Middle clouds
    (a.) altocumulus (Ac)
    (b.) altostratus (As)
    (c.) nimbostratus (Ns)
3. Low clouds
    (a.) stratocumulus (Sc)
    (b.) stratus (St)
    (c.) cumulus (Cu)
    (d.) cumulonimbus (Cb)

Cumulus and cumulonimbus clouds of vertical development often protrude like fingers in the middle and upper level of the troposphere and often arch through the upper-level clouds to pierce the tropopause. The nimbostratus, altostratus, and altocumulus — and often the nimbostratus — infiltrate other levels of the atmosphere. Altostratus can also be observed at higher levels. Stratocumulus and stratus are strictly low-level clouds. In addition to these ten cloud genera, there are numerous species and varieties of cloud which sometimes inhabit the sky. It is essential for pilots to know the weather associated with each. A wise pilot uses the clouds encountered along his route as signposts of air safety. That's what the game is all about!

2 ATMOSPHERIC HEAT

The primary source of energy utilized by the earth's atmosphere is the sun. This energy arrives in the form of short-wave radiation. Actually, the sun is the direct source of most earth energy forms. Even exceptions such as geothermal and nuclear energy remain indirect recipients of solar power.

The sun itself is a high-energy source that apparently obtains its energy from a complicated process of hydrogen-to-helium integration called *fusion*. Physicists have long been convinced that the final mass of the helium produced is slightly less than that of the original hydrogen atoms used in the fusion. What happens to the excess mass? Is it lost forever? No, according to Albert Einstein's special theory of relativity, the mass is converted into energy.

Let's try a simple mental Einstein-type experiment. Our goal will be to calculate the value of the intensity of solar radiation at *any* distance in space from the sun. First, let us assume that the sun is spherical in shape and has a radius, r. Next, we shall assume that the sun is at the center of a second and larger sphere of radius, R. The R will also represent the distance that a planet (for instance, earth) may be from its sun. Let's call the intensity of radiation emitted at the sun's surface I_s. The sun's surface area ($4\pi r^2$) may be included to develop an idea of total solar energy. Since we know that energy cannot be created or destroyed, energy radiated from the solar surface must equal the energy received on the interior surface of the make-believe sphere surrounding the sun.

Equating the energy radiated with the energy received yields the following mathematical expressions:

$$(4\pi r^2) I_s = (4\pi R^2) I_e \qquad (2\text{-}1)$$

and

$$I_e = (r/R)^2 I_s \qquad (2\text{-}2)$$

In the equations, I_e refers to the radiation intensity received at the surface of the larger sphere. If the larger sphere "just happens" to be the planet earth, and the radius of that sphere is assumed to be the average distance between earth and its sun, then I_e is also the valid *solar constant*. The latter is assumed to be a specific, nonchanging value. What does this imaginative experiment tell us? It shows that radiation intensity decreases as the square of the distance from the radiating sphere increases; in other words, we have just demonstrated an example of the *inverse square law*.

Solar energy is attenuated further by absorption and scattering as it penetrates our planet's atmospheric shield. Nicknamed "insolation" (a contraction for *in*coming *sol*ar radi*ation*), this is simply received energy from the sun.

HEAT AND TEMPERATURE

The terms "heat" and "temperature" are not interchangeable. Each describes a different phenomenon. According to the kinetic theory of matter, heat is simply a mode of molecular motion or energy. Matter is not, as once was thought, a solid homogeneous mass, but instead is composed of identifiable structural elements called *molecules*. Molecules are made up of even tinier elements called *atoms*. The molecules are like miniature solar systems and the velocities of their orbiting bodies represent kinetic energy. Heat can also be viewed as an energy form described by the motion of molecules and the atoms that make up the molecules. If a molecule is a "solar system," the world inside the atom must span galaxies!

Now, how much heat does anything contain? This is determined by the number and structure of its molecules and the sum of their individual kinetic energies. The latter is given by the expression $\tfrac{1}{2}mv^2$, where v is the mean linear velocity of the molecules, and m denotes each molecule's mass. So much for heat.

What is temperature? Simply the mean kinetic energy of the molecule. To a human organism, temperature also has a physiologic meaning, which in certain situations (flight extremes of heat and cold) can acquire great relevance. The billions of nerve receptors in our bodies react directly to temperature stimuli of all kinds. A nerve center, called an axon, has tentacle-like feelers of five variant types attached to it. These feelers, called dendrites, are connected through the body in vast networks, from cell to cell, filling synapse after synapse. Among the dendrites there are some that respond solely to temperature-related stimuli. Clusters of nerve cells congregate close to the skin surface (epidermis) in certain body areas, especially under the nose and on the palms, the soles of feet, and the genitals. Receptors in these particular areas are vulnerable to even small temperature fluctuations and will react most quickly.

Now, let's imagine a second experiment to aid our comprehension of temperature. Consider this situation: A glacier is being studied by two patient geologists. The glacier's mass is a billion kilograms; its mean temperature is -5°C (which is equal to 268°K on the Kelvin scale); its specific heat is 0.5. The scientists are working on location, that is, on the glacier. Each man weighs 70 kg, has a body temperature of 37°C (310°K on the Kelvin scale), with heat capacitances of close to unity. Would the scientists or the glacier contain more heat? Let's also assume that there is a specific temperature (or lack of temperature) at which all molecular activity ceases. This temperature (called *absolute zero*) is considered to be -273°C (0°K on the Kelvin scale). Let's make a further assumption. The total heat in any substance is a product of its mass, molecular structure, and the total number of degrees *above* absolute zero. So, the glacier's heat would be 10^9 kg $\times$ 0.5 $\times$ 268°K = 1.24 $\times$ 10^{11} units of heat. The scientists' total heat would approximate 2 $\times$ 70 kg $\times$ 1 $\times$ 310°K = 4.34 $\times$ 10^4 units of heat. The glacier would therefore contain 2.857 million times as much heat as the humans, even though their temperature is 43°K warmer! It is apparent that the heat content of matter depends on three particulars:

1. The average molecular energy, or *temperature*.
2. The total number of molecules, or *mass*.
3. The molecular structure, or *composition*.

Each can be an important factor.

Temperature can also be measured. How did this ability evolve? The history of measurement often begins with comparison; so it is with temperature. Comparison began with ideas like, "That fire is warmer than the river," or comments like,

17 Atmospheric Heat

"It's warmer than yesterday." The next series of questions to arise was not only obvious but also pertinent. How much warmer or cooler was it? Some people assert that the idea of temperature scales is actually ancient as opposed to the actual creation of the thermometer itself. The idea of temperature scales was probably first suggested by the Greek physician Galen. The evidence supporting this argument is plausible but as yet inconclusive. More tangible developments began with Galileo Galilei's thermometer; about 1596 he described an instrument for examining degrees of heat or cold. Next, a definition was needed to outline a temperature scale, one that included fixed reference points. Bartolo's posthumous contribution in 1679 mentioned that the temperature of snow and boiling water (H^2O) could be used for this very purpose. Later, in the eighteenth century, many scientists invented, developed, or devised various temperature scales for their thermometers (Figure 2-1). Certain temperature scales have withstood the passage of time better than others. D.F. Fahrenheit began making mercury-in-glass thermometers in 1717, and the scale he devised has traditionally been the most popular in English-speaking nations. The Fahrenheit scale defines 32° as the temperature at which ice melts and 212° as the boiling temperature of water at sea level; the two reference points thus have a range of 180°F.

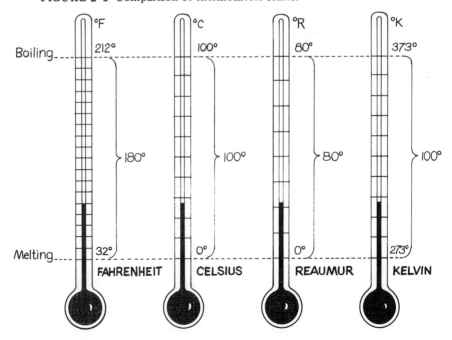

FIGURE 2-1 Comparison of thermometer scales.

The Reaumur scale, with reference points at 0° and 80°, was used extensively in central Europe until the late 1800s. In 1742, Anders Celsius, a Swedish astronomer, invented a temperature scale with a range of 100° between the melting point of ice and boiling point of water; the boiling point was to be determined when the atmospheric pressure was about 25 Swedish inches of mercury. This temperature scale is the most commonly used in the world today.

HEAT TRANSFER Heat can be transferred in various ways. One method is *conduction*. This term implies an energy transfer within or through a substance by means of molecular motion inside, with no help from the outside of the substance. Thus heat is transferred from molecule to molecule. Another mode of heat transfer within a fluid is *convection*, involving the movement of substantial volumes of the substance concerned. This process is of importance in the atmosphere, especially in the vertical exchange of heat and other air-mass properties. In reality, it is the physical transfer of a heated substance being moved to a new location. In meteorology, vertical motion is usually implied. Horizontal motion assumes the appellative of *advection*, although the process is similar. *Radiation* is the emission and propagation of waves transmitting energy through space or some other medium. In meteorology, the primary interest is in thermal radiation and the processes of radiational exchange. This exchange is governed by the absolute temperature of the radiating body.

There are other kinds of heat forms which are important to a pilot, or to aircraft operations in general. One is *latent heat*, or heat that cannot be sensed. Let's consider the treble phases of water: solid, liquid, and gaseous. Heat surely plays a role in the familiar phase transitions of melting, vaporizing, and sublimation, and these are vital in determining the atmospheric heat budget. When a phase change occurs, a transfer of heat also occurs, along with a change in volume. Only the temperature and pressure remain stable.

Why is latent heat important to aviation? How's this example? Power plant icing, sometimes called carburetor icing. The physics of this problem is demonstrated when the aircraft fuel is vaporized for combustion. To supply that energy required to vaporize the fuel, additional heat must be taken from the carburetor's environment. This process cools the carburetor to a temperature just below freezing. If the air happens to be moist, then ice forms in the air and fuel ducts inside. This, in turn, retards the flow of fuel and air to the combustion chamber, eventually leading to engine failure. To a pilot, this is hardly a source of merriment. This problem can easily be avoided simply by making sure the carburetor heater is in working order and using it properly.

Latent heat causes other problems in aviation. When an air bubble is buoyant, it rises and cools adiabatically; condensation occurs, releasing the latent heat, which then becomes sensible heat, which in turn warms the environment and makes the atmosphere buoyant. The buoyant air rises as convective cells, manifested as turbulence. When liquid water is present and the temperature is at or below freezing, structural icing may be encountered. This, together with turbulence, can cause serious problems for aviation. Yes, latent heat can be a bugaboo.

The primary warming mechanisms of the atmosphere are convection and long-wave radiation from the surface.

TEMPERATURE DISTRIBUTION

Atmospheric temperatures vary by latitude and altitude and often undergo abrupt changes season by season. The temperature of the air is also influenced by topography (physical characteristics of a given surface, such as local exposure, color, terrain cover, and so on).

Where did our "latitudes" come from? In the sixth century before Christ, Ulinderas, a student of Pythagoras, taught that the earth was round. About 500 B.C., Parmenides originated the concept of latitude: five climatic zones dividing the world. The Torrid Zone is far to the south of Greece, and here the sun is always high in the sky. This zone is bounded by two imaginary lines, the Tropic of Cancer to the north, the Tropic of Capricorn to the south. Beyond these boundaries lie the North and South Temperate Zones. An Arctic Zone completes each hemisphere in the extreme north and south.

Insolation affects the temperature, more so in the mountainous regions. Also, the various elevations of a mountain system are dominated by temperature regimes that may be related. Adiabatic cooling, for instance, one factor in determining the temperature range, makes mountain air cooler and thinner than flatland air at similar latitudes. Also, mountains act as heat sources and sinks that result from the various exposures and orientations of the terrain. These, in turn, warm and cool the adjacent air. The location of these sources and sinks are evident on micro- (small) and oftimes meso- (intermediate) scale climatic maps on which the temperature distributions are indicated by *isotherms* (lines of constant temperature). Hills and mountains create heat sources and sinks because of varying exposures to both insolation and outgoing long-wave radiation. Exposure determines the amount and intensity of insolation received by a sloping surface. K. Schutte in 1943 devised a simple procedure to determine the quantity of heating related to insolation received by a surface. He suggested that five factors are involved:

1. Latitude.
2. Solar declination (the season of the year).
3. Time of day.
4. Angle of slope.
5. Orientation of the slope.

Cloudiness and turbidity of the air are also important nonastronomical factors. When each of these factors is considered, it becomes obvious that mountains do indeed act as heat sources and sinks. In addition to these factors, we must also consider that mountains rise into reduced atmospheric densities. This means that a lesser amount of radiant energy is attenuated prior to being absorbed at the mountain surface, and more energy will be absorbed on the mountain slope than on the valley floor. Because of the thinner atmosphere on the surface of the mountain, a greater amount of the heat received during the hours of daylight is lost as long-wave radiation during the nighttime hours. The total effect is that mountain tops and slopes will heat and cool at a greater rate than does a valley floor.

TEMPERATURE LAPSE RATE AND ATMOSPHERIC STABILITY

One concern of every pilot is the stability of the aircraft. A stable aircraft will return to straight-and-level flight automatically if disturbed, but an unstable aircraft will continue to deviate from normal flight attitude.

The atmosphere also possesses various types of stability. The vertical stability of the atmosphere can be defined as the air's ability to return to its original position or level after being disturbed even by minor impulses. So, a stable atmosphere is one in which buoyancy forces oppose vertical displacement of the air from its original level. When the air is unstable, it tends to continue to move further away from its initial altitude or level after it has been disturbed. Buoyancy forces enhance its vertical motion. Neutral stability occurs when the buoyancy forces neither oppose nor enhance the initial disturbance.

To understand air as forces of buoyancy act on it, visualize an imaginary bubble of air encased in an elastic film. The film acts as a heat barrier, and the air bubble is assumed to be at the same temperature as its environment while at its initial level. When the bubble rises, its volume will expand as the pressure decreases. When it sinks, its volume will contract as the pressure increases. Since the elastic film covering the air bubble acts as a heat barrier, the encased air expands and contracts without gaining or losing heat from or to the surrounding air. This means that the system is *adiabatic*.

Now if we examine the relationship denoted in Equation 1-1, we see that as pressure increases the temperature increases, and vice versa. Thus, cooling within the bubble occurs because of the heat required to expand the bubble as it rises. Conversely, if the bubble descends, its volume contracts. Energy required to compress the air is converted to heat, and the air contained in the bubble increases in temperature as the environmental pressure acting on the bubble increases, as required in Equation 1-1. This is known as an adiabatic process, meaning that our bubble of air neither gains heat *from* nor loses heat *to* its environment. The dry-adiabatic lapse process is one without a moisture change phase, that is, no condensation, evaporation, fusion, or sublimation. The saturated-adiabatic process is one in which phase changes do occur and the latent heat is considered. Saturated air undergoes rapid pressure and temperature changes that cause a conversion of latent heat to sensible heat and vice versa. This process occurs whenever latent heat is converted into sensible heat, thereby raising the temperature of the bubble as condensation takes place. Water substance must be present in the atmosphere for this procedure to occur.

The degree of stability or instability of the atmosphere depends on the buoyancy forces present at any given time and place. These forces cause and sustain the atmosphere's vertical motions. Buoyancy was first recognized about 250 B.C. by a Greek mathematician and inventor, Archimedes.* His principle of buoyancy applies to all fluids, that is, gases and/or liquids: *"a body immersed wholly or partly in a fluid is buoyed by a force equal to the weight of the fluid displaced."* In the atmosphere, instead of dealing with bodies of solid matter immersed in a liquid, we become interested in air bubbles immersed in a fluid air (a thousand times less dense than water). To understand

---

*Archimedes (*c.* 287-212 B.C.) used the principle of buoyancy to determine the percentage of gold in the crown of Hiero, king of Syracuse.

the physics of this principle, we must develop a mathematical expression that can describe and define density. Imagine our bubble of air occupying a volume V, with density ρ' and weight $\rho'V'g$, where g is the acceleration of gravity. The weight of an equal volume of environmental air would be ρVg. Next, assume that the volumes of our imaginary bubble of air and the environmental air the bubble displaces are equal. The force that causes warm bubbles of air to rise and cold ones to sink is the buoyancy force. This can be stated mathematically:

$$a = g(\rho - \rho')/\rho' \qquad (2\text{-}3)$$

where a is the vertical acceleration. Since the density of the atmosphere cannot be measured directly, a substitution of density from Equation 1-1 is made, and we now can express density in terms of temperature, which can be measured:

$$a = g(T' - T)/T \qquad (2\text{-}4)$$

Note that ρ' and T' refer to the bubble, and ρ and T refer to the environment.

Equation 2-4 indicates that when the temperature of the bubble is warmer than the temperature of the environment (or $T' > T$), the vertical acceleration will be positive, upward motion; when $T' < T$, the motion will be negative, downward toward the surface.

Now let's examine the adiabatic process, since it is so important to vertical motion and the stability of the atmosphere. We'll need to use the first law of thermodynamics, which states that the change of a system (h) is equal to the internal energy (E) of the system plus the work done (W) by the system; it is written as

$$\partial h = \partial E + \partial W \qquad (2\text{-}5)$$

We will rewrite Equation 2-5 for a constant-pressure process, since it is easier to handle for the atmosphere in that form than the constant-volume method. Then, holding the pressure constant, $\partial E = c_p \partial T$ and $\partial W = -\alpha \partial T$. Since the process is adiabatic and no heat is gained or lost by the system, then $\partial h = 0$ and Equation 2-5 becomes

$$\partial h = 0 = c_p \partial T - \alpha \partial p \qquad (2\text{-}6)$$

Now, introducing the hydrostatic equation, because we are looking at hydrostatic equilibrium:

$$\alpha \partial p = - g \partial z \qquad (2\text{-}7)$$

Now, combining Equations 2-6 and 2-7, we can derive an equation from which the adiabatic lapse rate can be obtained:

$$0 = c_p \partial T + g \partial z \qquad (2\text{-}8)$$

By solving for $\partial T/\partial z$, the adiabatic lapse rate can be described:

$$\partial T/\partial z = - g/c_p \qquad (2\text{-}9)$$

Thus, for a dry atmosphere (nonsaturated) where $g = 9.8$ m s^{-1} and $c_{p\Delta} = 1{,}004$ m^2 s^{-2} °C^{-1}:

$$\partial T/\partial z = - 9.8/1{,}004 = - 0.98°C/100 \text{ meters} \qquad (2\text{-}10)$$

and for a saturated atmosphere where $c_{ps} = 1{,}911$ m^2 s^{-2} °C^{-1}:

$$\partial T/\partial z = -9.8/1{,}911 = - 0.51°C/100 \text{ meters} \qquad (2\text{-}11)$$

Now that the dry and saturated (moist) lapse rates have been defined in Equations 2-10 and 2-11, we can apply our knowledge of the adiabatic process to the imaginary bubble of air as it moves vertically. The bubble of air at its initial position (level) has a temperature equal to that of its environment. Since it is adiabatic, it will rise and fall with its temperature changing at the adiabatic lapse rate.

For example, if the bubble is not saturated and it is disturbed, it will rise or fall — and cool or warm — at the dry-adiabatic lapse rate of about 1°C/100 meters. If the environmental lapse rate is 1.2°C/100 meters and the bubble is displaced vertically from its initial level, then the bubble's temperature will always be warmer than the environment when sinking and cooler when rising. This type of environmental lapse rate produces a condition of *absolute instability*, because the buoyancy forces will always accelerate the bubble further away from its original position (level) in the atmosphere. Area A of Figure 2-2 illustrates this condition.

Another stability classification occurs when the environmental lapse rate is equal to the dry-adiabatic lapse rate for a

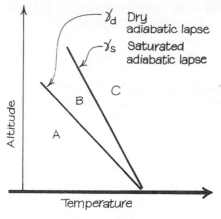

A - Absolutely unstable
B - Conditionally unstable
C - Absolutely stable

FIGURE 2-2 Atmospheric stability criteria.

dry atmosphere, or equal to the saturated-adiabatic lapse rate if the atmosphere is saturated. These conditions produce *neutral stability*, since no buoyancy acceleration would be exerted on the bubble of air and because the temperature of the bubble would be equal to that of the environment.

The *conditional stability* (or *conditional instability*) classification occurs when the atmosphere's lapse rate is somewhere between the dry-adiabatic and the saturated-adiabatic lapse rate. Conditional stability implies that conditions must be met in order to have stability or instability. For the atmosphere to be in stable equilibrium, it should not be saturated. This means that when the dry-adiabatic lapse rate of the bubble is compared to the environmental lapse rate, the air will be stable. When disturbed, a rising bubble will always be cooler than the environment — and warmer than the environment when it sinks. This means that it will always be accelerated toward its original position (level); by definition, this is stability. But if the air is saturated and has a saturated-adiabatic lapse rate, the bubble will be warmer than the environment when disturbed upward and cooler when disturbed downward. This, as in the first example, means that the bubble will always be accelerated away from the original position (level); by definition, this is instability. Area B of Figure 2-2 illustrates these criteria.

The final stability classification of interest is illustrated in Area C of Figure 2-2: If our imaginary air bubble is immersed in an atmosphere whose lapse rate is less than that of the saturated-adiabatic lapse rate, a condition of *absolute stability* exists. This can be demonstrated by applying the buoyancy principle: In this example, let's assume that an initial disturbance will cause the bubble either to rise or to sink. The rising bubble will be cooler and the sinking one will be warmer than the environment. The buoyancy forces under stable conditions always cause the bubble to return to its original altitude (level).

If one portion of the atmosphere is unstable, we cannot assume that the entire atmosphere is unstable as it is usually stable. Instability usually occurs in selected areas from the ground layer to the various layers above the ground. Usually, at higher altitudes nearer the tropopause, the environmental lapse rate approaches that of the dry-adiabatic lapse rate, and stability becomes neutral.

The stability (or instability) of the atmosphere indicates the amount of turbulence a pilot can expect to encounter. The pilot flying in a stable atmosphere will have smooth flying most of the time, as indicated by the presence of stratiform clouds. When the atmosphere is unstable, clouds will be of cumuliform shape. The latter situation can create many problems, and some can be quite serious. More will be discussed about instability and its resulting turbulence in Chapter 7.

3 ATMOSPHERIC PRESSURE AND ITS INFLUENCE

The discovery of atmospheric pressure by Torricelli was unique. Scientists (including Torricelli) had long been seeking the "ideal vacuum." This challenge had been born in response to ancient dogma: "The existence of a vacuum anywhere in the universe is impossible." The "logic" behind this theory was attributed to Aristotle: Light could not penetrate a vacuum since the sun, moon, and stars were visible. Hence, the existence of a vacuum was impossible! It seems ironic that one of the greatest philosophers could be guilty of such faulty reasoning. Later, the "vacuum search" was further hindered because scientists in Galileo's age didn't believe that the atmosphere possessed weight. It was during this era that many investigators were attempting to create a vacuum to prove Aristotle wrong. The common procedure was to use a long tube sealed at one end and fill it with liquid quicksilver (mercury). Next, the open end of the filled tube was covered, and the tube was inverted and submerged in a basin of residue fluid as the cover was removed. The experimenter speculated as to whether a vacuum was created in the tube as the quicksilver receded in the tube. Torricelli, in his experiment, used two tubes approximately 90 inches long, each filled with mercury. When the tubes were inverted and the ends submerged, Torricelli noted that the heights of the two mercury columns were identical. He soon realized that the weight of the column of

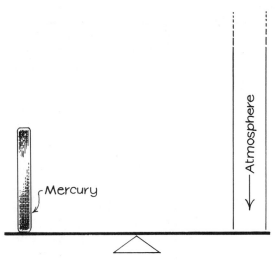

FIGURE 3-1 Principle of the barometer.

mercury was balanced by the weight of the atmosphere immediately above it. Thus Torricelli invented the barometer (Figure 3-1), and with it the concept of atmospheric pressure measurement. Torricelli was the first to recognize that the atmosphere exerted a pressure, and it was the torricellian tube that made possible science's attempt to find a relationship between those strange fluctuations of the height of the column of mercury in the tube and atmospheric change. Torricelli was the first to recognize that the mercury column fluctuations were caused by variations of the atmospheric pressure and that his instrument, the barometer, could be used to measure these variations.

Atmospheric pressure is very important in meteorology. It is the basic standard by which weather maps are analyzed. The density of the atmosphere at sea level is about a thousand times less than the density of water; thus, a column of mercury standing 29.92 inches high in a torricellian tube will balance the average weight, or pressure, of the atmosphere at sea level.

Another type of barometer using the same basic principle is the aneroid barometer. This instrument is an elastic type of pressure sensor. It uses a partially evacuated elastic chamber that contracts as the pressure increases and expands as the force of the atmospheric pressure acting on the chamber decreases. This chamber, called an aneroid cell (*aneroid* is a Greek word meaning without liquid), has a pointer attached that will indicate any movement (expansion and contraction). The movement of the pointer is calibrated to denote the pressure in millibars, inches of mercury, or millimeters of mercury. When an aneroid

TABLE 3-1 The Various Units Used to Denote Atmospheric Pressure

Pressure at sea level	=	1 atmosphere
	=	29.92 inches of mercury
	=	760 millimeters (mm) of mercury
	=	760 torrs
	=	1.01325 bars
	=	1,013.25 millibars (mb)
	=	$1,013 \times 10^2$ pascals (pa)
	=	1.01235×10^6 baryes

instrument is used as an altimeter, it will indicate altitude or elevation above sea level.

Atmospheric pressure may be expressed in many ways, by a variety of units. The particular units used depend on which country the pilot is flying from or to. Table 3-1 lists the various units. V. Bjerknes suggested that the term *bar* (a Greek word meaning weight) be used to define 1 million dynes per square centimeter of pressure. This term was accepted by the meteorological community and has come into general use throughout the world. But since the bar represents merely 1 million dynes per square centimeter, something more meaningful is necessary. Note that force is defined as the product of mass times acceleration. With this definition in mind, and knowing that the mass of the atmosphere is very difficult to determine, we must "juggle" the mass of the atmosphere to a more meaningful form. This can be done by using the specific mass or density for mass. We know that the standard atmosphere can balance a column of mercury 29.92 inches (760 mm) high. We also know that the density of mercury is 13.5955 g ml^{-1} (cm^{-3}), and the acceleration or 1.01325 bars. Most airports and weather stations, however, are at sea level can now be found to be the product of the density, gravity, and the height of the mercury column in centimeters (cm). This value is 1.01325×10^6 dynes per square centimeter or 1.01325 bars. Most airports and weather stations, however, are located at elevations above sea level. Thus, if the pressure were given in bars most station pressures would be given in decimal fractions! It is common knowledge that most people dislike fractional numbers, since they may cause some confusion. Therefore, millibars are used rather than bars for pressure denotation. If we use millibars instead of bars, standard sea-level pressure is expressed as 1,013 millibars (rounded off), and the millibar is the universal pressure unit used by meteorologists. The aviation community, especially in the United States, uses inches of mercury as the pressure unit for altimeter settings. This practice avoids creating the confusion of using sea-level pressure values for altimeter settings. The interchange of these values could lead to an error in the indicated altitude by the altimeter. This problem is caused by the use of different methods of converting station pressures measured above sea level to sea-level values. The altimeter setting assumes that the atmosphere is always in agreement with the ICAO (International Civil Aviation Organization) standard lapse rate, and indicates the altitude at which a given pressure will be observed. On the other hand, the sea-level conversion uses a theoretical reduction of the station pressure to sea-level pressure, and includes the 12-hour mean station pressure in the computation.

The altimeter setting value of the atmospheric pressure, to which the scale of a pressure altimeter is set, reflects the true altitude of the field elevation at the location for which the value was determined. The field elevation is the exact height above mean sea level. With a correct altimeter setting, the altimeter should read field elevation and *not* zero (sea level). At Miami International, for instance, the field elevation is 12 ft; with the correct altimeter setting, the altimeter *would* read 12 ft.

The vertical pressure structure of the atmosphere was discovered during an experiment conducted by Florin Perier of Clemont, France, on September 19, 1648. When results of Torricelli's famous experiment of 1643 had reached France, Blaise Pascal wrote to Perier, his brother-in-law, and suggested that he take a torricellian tube (barometer) up the Guy-de-Dôme to see if the height of the column of mercury would vary as the ascent was made. Perier performed this experiment and, for a control, left a second torricellian tube at the monastery in Clemont to be observed by one of the monks. Although the monk involved has been forgotten by time, the instrument at the monastery indicated a reading of 3½ lines (26 inches), while the torricellian tube taken to the top of the Guy-de-Dôme had an average value (after several readings) of 2 lines (23 inches). This demonstrated what Torricelli had hypothesized, "We live submerged at the bottom of an ocean of elementary air, which is known by incontestable experiments to have weight, and so much weight, that the heaviest part near the surface weighs about one four-hundredth as much as water."*

The discovery by Perier in 1648 brought to light an impor-

---

*Middleton, W.E.K. *Invention of Meteorological Instruments*. Baltimore, Md.: Johns Hopkins Press, 1969.

tant concept vital to both aviation and meteorology. Perier showed that comparing the atmospheric station pressures of airfields located at various elevations would be nearly impossible in an operational sense. An example of this can be seen by comparing the station pressures at Salt Lake City (SLC)—elevation, 4,227 feet; and Oakland (OAK)—elevation, 7 feet. On a given day the station pressure at SLC might be 868 mb, while the value at OAK might be 1,017 mb. At other airfields, the pressures might vary from 1,030 mb for airfields near sea level to 730 mb for airfield stations at high elevations in the Rocky Mountains. Thus, airfields located at different elevations will have dissimilar station pressures, which in turn will reflect merely elevation differences rather than changes in the horizontal atmospheric pressure systems, otherwise known as "weather." To alleviate this difficulty, the meteorological and aviation community developed a method by which station pressures could be normalized to a standard datum plane. The most convenient reference level for most purposes is sea level. The station pressure is reduced to the value that would exist at a point at sea level directly below if the air of a temperature corresponding to that actually present at the surface were present all of the way down to sea level. (In practice, this reduction is made using a mean temperature for the preceding 12 hours.) In addition, a plateau correction is used for stations located at higher elevations. This means that sea-level pressures are valid for synoptic analysis, and the forecaster can therefore predict the movements of storms and the weather associated with them. These aspects of weather will be discussed in greater detail in subsequent chapters.

We've indicated how important atmospheric pressure is to meteorology. It's important in a very special way to aviation. No plane flies without an altimeter to tell the pilot how high the plane is, and altimeters are "set" according to atmospheric pressure. Altimeter settings are defined to provide standard altitudes for aircraft while in flight. In order to do this, a vertical temperature constant distribution (VTCD) of the atmosphere must be assumed. The VTCD used is the standard ICAO Standard Atmosphere. This means that each barometer (altimeter) is constructed to indicate the pressure-height of the ICAO atmosphere. This altimeter standardization provides the entire aviation community with improved air safety, since all aircraft will use the same reference base for their indicated altitudes, and space separation can be maintained for all planes flying in the various directions.

From the gas law discussed in Chapter 1, we discovered that the pressure exerted by a gas is proportional to the product of its density, pressure, and temperature, with the gas constant (R_d) being the constant of proportionality:

$$P = \rho R_d T \qquad (3\text{-}1)$$

where ρ is the density of the air, P is pressure, and T is the absolute temperature. The atmospheric density is the ratio of the mass per unit of volume. It should be noted that it is a tricky proposition to measure either the mass or the volume of the atmosphere, but atmospheric density can be substituted for the mass in the force or pressure equation. The value of atmospheric density can be determined from Equation 3-1, since both temperature and pressure values are available at all weather stations.

Remember that Perier's experiment in 1648 demonstrated that as elevation or altitude increases, pressure decreases, and vice versa. This is true not only in the atmosphere but in any fluid. These changes of pressure and height are opposite, or negative, in direction of action; therefore, it is necessary to use the negative sign when using this relationship. The law that Perier demonstrated in his famous experiment is the law of hydrostatic equilibrium, and can be written in precise mathematical form. To do this, first force and its terms must be defined: Mass is m, acceleration is a (which, in the case of the atmosphere, is equal to g, gravitational acceleration), and area is S. Force, as defined by Newton, is the product of mass and acceleration:

$$F = ma \qquad (3\text{-}2)$$

and pressure is force per unit area:

$$P = F/S \qquad (3\text{-}3)$$

or energy per unit of volume:

$$P = Fz/\text{volume} \qquad (3\text{-}4)$$

Finally, weight, or gravitational force, is

$$W = mg \qquad (3\text{-}5)$$

Since we have defined pressure in Equation 3-4 as energy per unit of volume, a change in the pressure with respect to a change in elevation is now written

$$\partial P = -\rho g \, \partial z \qquad (3\text{-}6)$$

The hydrostatic equation demonstrates that the change in pressure (∂P) of a column of fluid equals the product of the density, gravity, and height change (∂z) of the column (the atmosphere in this case). If the density of the fluid (air) from Equation 3-1 is substituted into Equation 3-6, the hypsometric equation is obtained, and this is given in its differential form in Equation 3-7:

$$\partial P = -Pg\partial z/R_d T \qquad (3-7)$$

Upon integration (a higher mathematical series of manipulations) Equation 3-7 becomes

$$P = P_0 \exp{-[g(z - z_0)/R_d \overline{T}]}$$

Equation 3-7 describes the pressure as a function of the height and mean temperature of the atmosphere. This relationship is used to compute the heights of the various pressure levels of the atmosphere. It is also used in determining the pressure-height relationship utilized by the altimeter. These values are determined in the following manner: First, the standard lapse rate of the ICAO atmosphere is assumed. Then the pressure is computed to height interval by dividing the atmosphere into successive layers from the ground up. For each layer, a mean temperature for the ICAO lapse rate for that layer is determined. Then, using the mean temperature value that has been found, the pressure-height relationship is obtained. This pressure-height relationship is important as a safety tool for flying. All pilots have discovered for themselves that the pressure decreases as the altitude increases. Another case of pressure variation is a temperature-volume relationship discovered by physicists Charles and Gay-Lussac. They recognized that (nearly all) matter expands when heated (including air). By showing that when volume is held constant and temperature increases, the pressure also increases, they provided some valuable information. A simple demonstration of this can be provided by visualizing three balloons, another mental experiment. Each balloon is inflated to a common pressure. The first balloon is then cooled to a colder temperature than the temperature of inflation; the second balloon remains at the inflation temperature; and the third is warmed to a higher temperature. The result is that the first balloon occupies the smallest volume and the third balloon has largest volume.

In the real atmosphere, each of these balloons represents a column of air with a specific mean temperature different from the others. Since the amount of air in each of them is equal, we shall assume that the total mass of the air in each is the same; then the pressure exerted by each column against the surface upon which it rests will also be equal. It should also be noted that the pressures at the top surface of each column are equal, but not necessarily the same as that at the bottom surface of each column.

Remember that in an aircraft, height is measured with a pressure altimeter. This instrument is an aneroid barometer with its dial graduated in units of height rather than units of pressure. Let us assume that we are flying from Salt Lake City (SLC) to Oakland (OAK), and assume that we are flying at an assigned altitude. The assigned altitude is *not* the true altitude, since pressure altimeters are being used. This means that if the SLC altimeter setting is used along the entire route, the flight is actually being made along a surface of constant pressure (or isobar) relative to SLC. Thus anytime the mean temperature of the layer of air between the plane and the surface *increases* or *decreases*, the height of the pressure surface on which the plane is flying will also increase or decrease, as did the volume of air in the balloons.

Now, let us examine *how* the change in temperature of the air layer will affect the altitude of the aircraft. If the temperature were to decrease enroute, the plane would be flying lower than the indicated altimeter reading; similarly, if the pressure were to become lower, our true altitude would also be lower than the indicated altimeter reading. Obviously, if the temperature or pressure were to rise along the route of flight, the plane would be higher than the indicated altimeter reading. Thus, we can directly observe how a variation of either the horizontal temperature or pressure affects the flight of an aircraft with respect to altimeter indication. (We avoid the problems — and potential dangers — by correcting our altimeter settings with updated weather data broadcast by each FSS along our route of flight.)

Temperature also affects the flight characteristics of an aircraft through modification of atmospheric density. Remember, density is directly proportional to the pressure and inversely proportional to the temperature. Flight aerodynamics require that a given mass of air flow past the wings for the plane to become airborne and fly. Since the mass of the atmosphere is difficult to determine, we use the specific mass or density of the atmosphere. We are already familiar with density and how it is determined, so we shall not discuss it further. What it boils down to is that if a given plane requires 1,000 ft of runway to take off at sea level (at "normal" temperature, usually considered 59°F, or 15°C), the same plane will need a runway of about double that length for the takeoff run at an elevation of 5,000 ft! At 8,000 feet? Just triple the length of runway! And how long a runway would be required if the air was warmed to 95°F at a 5,000-ft elevation? That's another situation entirely. The density of the air at 95°F at 5,000 ft would be more nearly equal to the density of the air normally found at the 8,000-ft elevation at 59°F! Translated, this means that a pilot flying at 5,000 ft would be flying a plane that thinks (and behaves as though) it is flying at 8,000 ft; it will be much more sluggish in its performance. This phenomenon is known as *density altitude* and it is something that a pilot must be aware of whenever flying in and out of airfields at higher elevations. Pilots who choose to ignore density altitude may provide the next day's headlines in local newspapers they won't be around to read.

Now let us examine sea-level pressure patterns. Remember that to compare station pressures, we must first normalize them to the same level. Obviously, the pressure measurements from several stations must be taken at the same time. If we are to analyze and draw conclusions about pressure patterns and their effects on weather situations at any particular location, this step is essential. For this reason, weather data are first observed, then collected and analyzed.

The surface analysis enables us to study the pressure on a single surface, so we may understand two-dimensional pressure systems. A three-dimensional system is achieved by using rawinsonde observations on constant-pressure charts. These are

TABLE 3-2 The Pressure-Height Relationship in a Standard Atmosphere

PRESSURE (millibars)	Height	
	Meters	Feet
1,000	111	364
850	1,457	4,781
700	3,012	9,882
500	5,574	18,289
400	7,185	23,574
300	9,164	30,065
250	10,636	33,999
200	11,784	38,662
150	13,608	44,647
100	16,180	53,083

constructed both manually and by computer for various selected pressure surfaces (see Table 3-2).

Let's assume that you either are a pilot or aspire to be one. Are you interested in pressure variations and how they can affect your flight and safety? You should be. Here are some common causes of variation in atmospheric pressure:

1. The diurnal pressure variation. This has a natural period of 12 hours with maximas near 10 A.M. and 10 P.M. local time, and the minimas about 4 A.M. and 4 P.M. Diurnal fluctuations should be considered whenever using pressure tendencies.
2. The movement of pressure systems, as they migrate around the periphery of the polar front. (Yes, they do move.)
3. The changes in the intensity of the pressure system, whether it deepens or fills.

Next you might ask, what is a pressure system? A pressure system is either the vertical or horizontal pattern of pressure distribution. First, let's discuss the vertical pressure distributions. These are closely related to the distribution of atmospheric temperature and are defined thusly:

warm-core low (cyclone) A pressure area in which the central pressure is lower than any surrounding pressures. If its temperature is highest at the center, the low will decrease in intensity aloft.

cold-core low (cyclone) A center of a low pressure surrounded by higher pressures. When the temperature is the lowest at the center, this kind of low pressure will intensify aloft. As a cold-core low is present, the low pressure will extend aloft and may even retain a closed isobar at the 200-mb level. This kind of low-pressure area is usually very slow in migrating. In other words, it isn't a speedy mover.

warm-core high (anticyclone) A pressure pattern characterized by a high-pressure center surrounded by lower pressures. The fact that the temperature is warmer in the center than at the periphery means that the high intensifies aloft.

cold-core high (anticyclone) A high-pressure pattern characterized by colder temperatures at the center than at the periphery.

Driving forces in atmospheric circulation result from the

rotation of the earth and the pressure gradients that result from uneven horizontal distribution of insolation. Causes of variation within insolation regimes include: (1) cloudiness in all forms, (2) exposure of topography, and (3) different latitudes. This last item can be demonstrated. The equator receives extremely large amounts of solar energy, while the poles receive trivial amounts of that energy. The equatorial air rises as it is heated, and a low-pressure area of converging winds is found at the surface. These areas near the earth's midpoint are called the doldrums, and are characterized by rising air and lots of rain. The lesser amount of solar heating at the poles causes the air to subside, forming a heat sink at the surface with a high-pressure region as a result. Because the earth rotates on its axis every day, a dynamic region of high pressure develops at about 30° latitude both north and south of the equator, with a balancing low-pressure region located at about the 60° latitudes. These are global pressure features that form between the equator and the poles.

Air moves from areas of high pressure to those of low. This motion is similar to that of water, which also flows from regions of high pressure to areas of low pressure. In simple terms, it means moving from where it is to where it isn't. If the earth were not rotating, the wind would blow from the high pressures at the poles to the low pressure at the equator. Thus, all surface winds would blow from the poles, with a counterflow aloft occurring from the equator to the poles. In the Northern Hemisphere, the surface flow would be from the north toward the equator, with a countering southerly stream aloft. George Hadley, a British maritime lawyer, reasoned in 1735 that the air close to the equator warmed more quickly and to a higher value than the polar air, because the insolation received at the equator greatly exceeded that at the poles, making the air warmer at the equator than that of the cooler polar latitudes; thus, warm tropical air would rise and cool arctic air would tend to settle. He then calculated that moist warm air cooled adiabatically, traveling skyward and moving poleward. Upon arrival at the poleward terminus of its travels, it would sink earthward, creating a mass of high pressure at the pole. Conversely, the air at the surface would flow from the polar regions, or high latitudes, toward the equator. But Hadley wasn't finished: He recognized the consequences of the earth's rotation on moving air. Even though he couldn't understand that the wind was being deflected to the right in the Northern Hemisphere, he did recognize the apparent cause of the trade winds. The thermally derived low pressure at the equator and the high

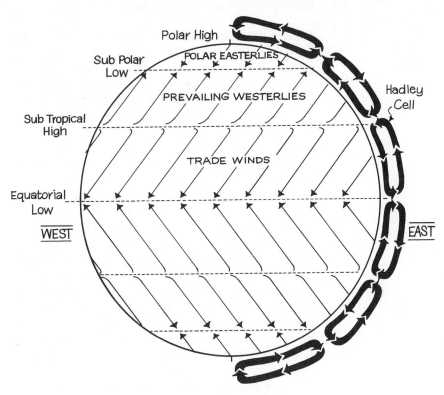

FIGURE 3-2 General circulation of atmosphere.

pressure at the poles combined with the steady trade winds became the "Hadley cell," named in his honor (see Figure 3-2).

Today, the general circulation models confirm Hadley's hypothesis of the thermally heated low-pressure band near the equator. Since Hadley's time, Coriolis has demonstrated the mathematics and physics of the earth's rotation. The "Coriolis force" pushes the winds toward the right in the Northern Hemisphere and toward the left in the Southern Hemisphere. What does the earth's general circulation resemble in other respects? There is high pressure in the polar regions. Winds originating in these regions are deflected toward the west, while most midlatitude winds blow from the west. A dynamic high is located at about 30° latitude in both hemispheres. A subpolar dynamic low is found near 60° latitude, and this is a famous spawning ground of migratory cyclones. Introducing the dynamic high and low pressure areas greatly improves the Hadley model of general circulation and makes the circulation more representative of the actual wind and pressure patterns found on the earth.

The polar winds are named the polar easterlies. The subtropical winds are the tropical easterlies and assigned the common name "trade winds." The remaining wind system in each hemisphere blows from the subtropical high-pressure region poleward to the subpolar low-pressure region of the high latitudes.

To summarize the general circulation pattern, we find that because of the eastward rotation of the earth, air particles moving either toward the north or south will possess an eastward component of motion, which is equal to the linear speed of the earth's surface at the latitude in which the motion was initiated. When the air motion is poleward, the eastward component of the wind velocity will increase and move faster than the linear component of the earth's surface at the higher latitudes. So the air is deflected toward the east, and the wind system in the poleward-moving air will be westerly. An example of this is the prevailing westerlies of the general circulation. If the air is blowing toward the equator, then the air motion is moving away from the poles and its eastward component is less than the eastward component at the lower latitudes. The resulting wind is deflected toward the west. Thus, an easterly air motion results. The polar easterlies and the trade winds are good examples of this pattern in the general circulation. The trade winds tend to be steadier than either the polar easterlies or the prevailing westerlies.

The subpolar low-pressure region is the spawning ground of migratory cyclones that race about the hemisphere, tracing the perimeter of the *polar front*. These can be cooler or warmer than the surfaces over which they flow. The waves that develop on the polar front modify the storm paths around the periphery of the polar front in the hemisphere. These migratory low-pressure systems (cyclones) contain the midlatitude weather systems. The wind systems associated with and governed by the various pressure patterns should be understood by all pilots. Such an understanding will assist them in learning to fly *with* the weather rather than *against* it, to take advantage of the weather rather than falling victim to it.

Weather generally associated with an anticyclone is most often clear skies and fair weather in rural areas, but stagnant conditions with fog, smoke, smog, haze, etc., may result in and around urban and industrial locales. High pressure can be and usually is the culprit. These anticyclones are famous for subsiding air, forming temperature inversions that hold atmospheric aerosols close to the ground under the apex of the inversions. A comprehension of the wind and pressure is necessary for safe flight. Wind direction with regard to high- and low-pressure systems is most interesting because of the Coriolis effect. This force causes the wind to blow in a clockwise motion around anticyclones and counterclockwise about cyclones. This introduces Buy-Ballots law, a rule of thumb that is essential for people who invade the sky: "If, in the Northern Hemisphere, you stand with your back to the wind, the high pressure is on your right and the low pressure is on your left; in the Southern Hemisphere, the high pressure is on your left and the low pressure is on your right."

The general circulation pattern that has been described provides a description of the air flow at the surface of the earth. This pattern would be very accurate if the planet had a uniform homogeneous surface, but since 75 percent of the surface is covered by water, this just isn't the case. What tends to modify the general circulation pattern? Let's answer that query with an example. The Eurasian (Europe and Asia) land mass experiences intense warming during the summer months and intense cooling during the period of low sun (winter). It generates and superimposes its own land-sea circulation system (Asian monsoons) into the general circulation pattern of the earth. Because of the large land mass, and its intense warming, the air over this mass obeys the gas laws. When the warming occurs, the heated air rises, and an area of extreme convergence is created. The air over the land mass warms much faster than that over the adjacent oceans, leaving the oceanic area with higher pressure than is found over the continent. When a large gradient is generated, its own continental-sized wind system becomes a reality. Now, what forces drive and guide the cyclonic systems that provide our weather?

1. The horizontal pressure-gradient force, which is proportional to the pressure differences between adjacent locations. This force acts from high to low pressure and causes the air to flow and the wind to blow.
2. The Coriolis force, which acts opposite to the pressure-gradient force and results from the earth's rotation.
3. The centrifugal force, which develops from the curvature of the flow pattern over the earth's surface. It can be observed by examining the curvature of the isobars indicated on the weather charts.
4. The frictional force, which is a function of the roughness of the surface. This acts as an air drag; it also tends to slow the wind speed and cause the wind direction to be deflected toward low pressure.
5. Gravity, which allows less-dense air to lie over dense, heavy air in a stable atmosphere. It is also an important contributor to atmospheric pressure.

If the pressure-gradient force ($-\alpha\, \partial p/\partial n$) is equated to the Coriolis force ($2\Omega c \sin \phi$), the balance is defined as

$$-\alpha\, \partial p/\partial n = 2\Omega c \sin \phi \qquad (3\text{-}8)$$

This wind will travel parallel to straight isobars, because the Coriolis force is equal to and acts opposite in direction to the pressure-gradient force. Now, if we balance the pressure-gradient force against the sum of the Coriolis and centrifugal (c^2/r) forces, the result is the gradient wind equation, which describes the gradient wind balance:

$$-\alpha\, \partial p/\partial n - 2\Omega c \sin \phi \pm c^2/r = 0 \qquad (3\text{-}9)$$

These various forces describe the winds as they blow in the atmosphere, especially in the middle latitudes. The friction term is usually neglected, if the wind height is above the friction level (600 m or 2,000 ft). Friction, however, plays another role. As it weakens the gradient flow, it causes a slight cross-isobaric exchange of air near the surface. The circulation of the free atmosphere above the friction level is shown in Figure 3-3. It should be noted that the isobars are not straight (as the geostrophic assumption requires) but curved; of course, the gradient wind system describes the circulation better. This is a picture of the real wind system sure to be encountered by pilots.

Another category of winds should be mentioned: These are the local winds. They, like the monsoons of Asia, result from uneven heating of land and land-water surfaces, that is, mountain-valley, land-sea, etc. They are of thermal origin, somewhat similar to the air circulating around a fire. Cool air is drawn into the warm area, and the fire heats the air, causing it to rise. This, in turn, draws more surrounding cool air into the fire to replace the rising warm air, and an area of convergence is initiated. Thus, a microcirculation regime is born. The same physical process forms the land-sea breeze. During the day, the ground heats to a warmer temperature than the water surface, and a circulation pattern is initiated. At night, this pattern is reversed. A land surface is cooled by radiational cooling at a higher rate than a water body, and, as in the monsoons of Asia, the direction of flow reverses itself for the same reasons. Another variety of local wind is the mountain-valley wind system. This, too, is a thermal system, with the mountain tops warming much faster than the valley during the daytime, and cooling more rapidly at night. It should be mentioned that most local winds are classified as either warm

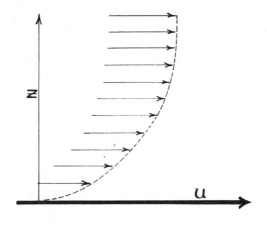
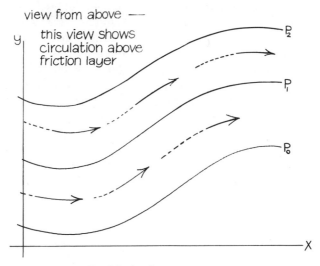

FIGURE 3-3 The friction layer.

or cold. Since wind is one of the most obvious manifestations of weather phenomena, there are hundreds of local names given to winds throughout the world. Air that is highly heated over the Sahara desert travels to the coastal areas of Africa, bringing hot, dry scorching conditions. This wind is called *sirocco*. If this wind reaches Spain, it is called *leveche*, and in the Canary Islands *leste*. In Australia, the hot northerly wind is called *brickfleder*. The warm, dynamically heated wind of the Southern Alps is called the *foehn;* its counterpart of the high plains of North America is given the Indian name *chinook;* the same wind is called

Santa Ana in southern California. These dynamically heated winds are generated when air is forced over a mountain and is adiabatically heated as it is forced to descend down mountain slopes onto the valley floors. The chinooks on the eastern slopes of the Rockies have been known to have warmed the air as much as 40°F during a 1-hour period. The Santa Ana winds of southern California have been known to have temperatures around 65°F (give or take a few degrees) and dew-point temperatures of –10°F! Other locations on the lee side of mountains have reported similar dynamically heated winds, such as the *Zonda* of the Argentine and the *koembang* of Java.

Then there are the cold winds. One example is the *norther* of the great plains of North America. This cold wind is imported from the Canadian regions of the Arctic. In a southward extension of the polar or arctic air mass that inhabits the polar regions, it moves southward into the plains and Mississippi Valley of the United States, bringing a tremendous and very rapid cascade of plummeting temperatures. Temperature drops of as much as 55°F in 3 hours have been recorded. Most regions have their own local names for cold outbreaks, such as the *buran* of Siberia, the *poorga* of Russia, the *friagem* in the Amazon valley, the *hahoob* in Egypt and the Sudan, and the *papagayo* and *tenuantepecer* in Mexico and Central America. Still another cold wind is the *bora*, a violent northeasterly wind of the Adriatic. This type of wind occurs when the cold air from highland areas pours into an adjoining lowland area with the assistance of a favorable pressure distribution. This type of wind may develop especially well on the edges of Greenland and Antarctica, where it reaches velocities to 60 knots and higher and may last for several days. It is called the *willywaw* in Alaska and at the Strait of Magellan. The weather associated with local winds can be a friend or foe to the pilot. With any wind, you can always expect to encounter turbulence, a hefty companion of air borne with shifty, gusty conditions at the surface. Any pilot knows well the hazardous circumstances of aircraft operations under turbulent conditions. Even taxiing airplanes can be difficult!

4 MOISTURE, CLOUDS, PRECIPITATION, AND OTHER HAZARDS

Atmospheric moisture creates and is the primary ingredient of weather. With the "wet," a variety of attendant weather hazards, especially for aviation, soon appear. If moisture were not present in the atmosphere, the only weather hazards would be turbulence, wind, and blowing phenomena. You may ask, if atmospheric water vapor amounts to less than 4 percent of the atmosphere's total volume, then why is it so vital to the weather process? To understand the moisture contribution, we should first examine the weather and moisture relationship. Also, what special properties does moisture have?

We have mentioned that ordinary water substance freezes (0°C, 32°F), boils (100°C, 212°F), and exists simultaneously in three forms: solid (ice), liquid (water), and vapor (gaseous). But it must be remembered that any form of water substance remains in the atmosphere for only a very short period of time. Water in the atmosphere is constantly in a transitional state between evaporation and precipitation.

Atmospheric moisture makes the most important contributions to the weather "worries" that pilots encounter. We mentioned earlier that the water vapor content of the atmosphere varied from nil to about 4 percent of the total volume of the atmosphere. Yet, this minor constituent is responsible for weather as we know it.

When water vapor condenses, it becomes one of the "meteors" from which the Greek word *meteorology* originates. The meteors of concern to meteorologists are: hydrometeors, lithometeors, photometeors, and electrometeors.

Hydrometeors consist of liquid or solid water particles that are either falling through or are suspended in the atmosphere, blown from the surface by winds, or deposited by winds on objects.

Lithometeors consist of a visible concentration of mostly solid, dry particles. The particles are more or less suspended in the air or lifted from the ground by winds.

Photometeors are luminous phenomena produced by reflection, defraction, diffraction, or interference of light.

Electrometeors are components of atmospheric electricity that can be seen or heard.

For those who challenge the skies, an understanding of hydrometeors and clouds is vital. These phenomena, formed by the condensation of water vapor, provide a potpourri of challenges to both safety and sanity. Let's explore the processes of atmospheric water substance, and discuss briefly condensation and precipitation. To begin with, the amount of water vapor that the atmosphere retains is a function of its temperature. As a general rule, the higher the temperature, the greater amount of moisture the atmosphere can retain. This means that if the relative humidity of the atmosphere is 50 percent at temperature T_a, but becomes 100 percent when the temperature is lowered to T_b, then for condensation to take place it would be necessary to have something for the vapor to *condense on*, and we call this thing a condensation nucleus. Before we discuss the role of condensation nuclei, we should examine the methods by which the atmosphere is cooled. In the case of clouds, cooling is by adiabatic expansion. Let's explain that. We know that when a tire is inflated, its stem is heated by the air as it is compressed and enters the tire; on the other hand, if the compressed air is released from the tire, the valve stem is cooled. And when expanding air strikes your hand, it feels cool. This same process cools moist, rising air, which then condenses and forms clouds. The air rises because of orographic or thermal lifting, which occurs when the earth's surface is heated along with the adjacent air.

The condensation process is most interesting. In order for it to occur, the initiation of either phase change from vapor to liquid (water) or vapor to solid (ice) must occur. The normal process in the troposphere is one of *heterogeneous nucleation*, in which the phase change is initiated by condensation nuclei, or *kernels*. If the kernels are not present in sufficient numbers to initiate the condensation process, supersaturation may occur. This happens when the atmosphere at a particular location contains a greater amount of water vapor than it would normally accommodate at a given temperature. Under average conditions, the atmosphere contains enough kernels for condensation to occur at 100 percent relative humidity. The kernels enter the atmosphere by natural and anthogenic processes. Later, they are dispersed by the winds, dust storms, sea salt spray, and combustion. Freezing water particles occur spontaneously (without aid of nuclei) at or below a temperature of about −40°C; this is called *homogeneous nucleation*. Condensation trails occur when an aircraft penetrates a layer of supersaturated air in the high atmosphere. Condensation may occur when relative humidities are as low as 95 percent. When water vapor condenses or liquid water freezes, the *thermodynamic process* must be considered. This involves the latent heat that is released to the atmosphere. The atmosphere loses latent heat when water is evaporated or ice melts.

The heat of vaporization is called latent heat because it cannot be detected by human senses. Sensible heat *can* be detected by our senses (felt). Latent heat is converted to sensible heat, which is then released by condensation to warm the atmosphere. This energy drives thunderstorms and generates severe weather. How does latent heat become so awesome, the force behind even the tornado? First, the water vapor condenses into small water droplets. These form clouds and fog if they survive and remain suspended in the atmosphere for a considerable period of time. Droplet clusters are called clouds. Clouds grow by sublimation, condensation, and coalescence. The large droplets consume the smaller droplets as the larger grow. This happens when a large droplet is next to a smaller droplet. The net flow of vapor will be from the smaller to the larger. Thus, the larger grow at the expense of the smaller. The larger droplet may continue to grow until no more water is available to support continued growth. Or it may continue to grow by coalescence, a term used to denote the growth of water drops by collision.

Something similar happens in the generation of snowflakes. Both ice crystals and snowflakes have vapor pressures lower than water drops; thus, when water and ice are both present and in close proximity to each other, the ice crystal or snowflake will grow at the expense of the water droplet, and the precipitation, if or when any occurs, will fall in the form of snow. All salts in solution (brine) cause a reduction of vapor pressure from that of pure water. This means that a salt nucleus is hygroscopic and a very good condensation kernel.

The presence of condensation nuclei is a very important part of the precipitation process. In fact, these nuclei can be created artificially to modify clouds and cause precipitation. This is useful because fog can be removed from airports by inducing clouds to precipitate if the temperature approximates a critical reading of -4°C (26°F).

Water droplets form clouds or fog. Clouds are above the surface. Fog isn't. Clouds and fog cause many problems to air safety. Either reduces visibilities or ceilings, thus endangering everything airborne. It is often necessary to restrict the use of visual flight rules (VFR) at air fields because of low ceilings and poor visibilities resulting from stratus clouds and fog.

What about droplet size associated with the various types of precipitation? Rain is the most common form of precipitation. Raindrops can be either in the form of large (0.02 inch, 0.5 mm) or small drops that are widely separated. Drizzle, on the other hand, appears to float while following air currents; unlike fog droplets, however, drizzle falls to the ground. The most dangerous precipitation encountered in flight is freezing rain or drizzle. These two freeze on impact with a surface, whether that surface is an aircraft in flight or the ground.

Fog and cloud particles are much smaller than either drizzle or rain. These particles are so small that they exhibit a very small fall rate and remain suspended in the atmosphere until drier air comes breezing in. Drizzle droplets are also small and have a tendency to fall in a floating manner through the air, because the terminal velocity of the falling droplet is directly proportional to the square of their radii. Thus the larger the drop (droplet), the faster they fall. These particles fall at rates of less than 700 ft per minute. Raindrops may fall at rates between 700 and 1,000 ft per minute. Shower particles, which are very large, fall much more quickly, at near 2,000 ft per minute. These shower particles are born of cumuliform clouds.

Another interesting fact about ordinary water substance is that while ice melts at 0°C, at temperatures between 0° and -40°C both water and ice can coexist. The -40°C temperature is called the Schaffer point, named after Dr. Vincent Schaffer, who suggested that liquid water cannot exist at any temperature lower than this value. He also once stated that when water reaches this critical temperature, clusters of water molecules may take up icelike configurations and grow in size sufficient for them to act as ice nuclei, upon which ice crystals may rapidly form.

This information has made possible fog and stratus dispersal operations in those areas where undercooled fog and stratus occur. These undercooled cloud and fog formations consist of liquid water droplets at temperatures near -4°C (26°F). Theoretically, then, if a method can be found to initiate freezing of cloud droplets, then fog should be dispersed. Dr. Schaffer conducted an experiment in a cloud chamber. He created an undercooled cloud at a temperature of about -4°C before introducing dry ice particles into the fog. Dry ice particles have a dry crystalline structure that is very similar to that of frozen water particles (ice crystals). The "natural" cloud droplets reacted with the dry ice crystals as though they were ordinary ice crystals. Immediately, a net transfer of water from the liquid to the crystal began, and the dry ice crystals grew into small snowflakes and precipitated to the surface. At locations where undercooled fog or stratus clouds occur, Schaffer's technique will reduce the fog sufficiently to permit an airport to become operational very quickly.

Let's examine and classify some of the types of precipitation that are of interest to the pilot. Precipitation can be (1) liquid, (2) freezing, and (3) solid. Freezing rain and drizzle are the most dangerous. Neither contributes to aircraft pilot longevity. Either of these can "ice" an aircraft and put it on the ground in a matter of minutes, if precautions are not taken. The wisest precaution is to avoid the icing conditions anyway, anyhow.

Solid precipitation comes in many forms; snow, sleet, graupel, and hail are examples of this type, and these often cause the greatest concern to the aviation community. Sleet is composed of transparent, hard, spherical pellets that can be rather noisy when they strike the skin of an aircraft. Hail is very dangerous because of its size. Hailstones may be as small as 0.25 inch (about 6 mm), but are often as large as several inches. If a vulnerable aircraft becomes entangled with a hailstorm, it may sustain damage ranging from a dented airframe or starred windows to serious structural damage. If you are gliding in a hang glider, the consequences may not be a joking matter. Jet aircraft in flight have often reported that their huge engines have ingested hailstones that damage the turboblades and can even drown the flame. Hail is a manifestation of thunderstorm activity and is most common to areas east of the Rockies on the High Plains and around the Mississippi Valley.

Clouds have been observed and used for making instant weather forecasts since the time that man first became aware of clouds. If we are going to discuss clouds, first they must be defined, and then perhaps we can better understand what we are dealing with.

A cloud is a visible group of tiny particles of water, ice, or both in the free atmosphere. This ensemble may include other particles of differing sizes such as those found in fumes, smoke, or dust.

The appearance of a cloud will depend on its nature, whether it is composed of water droplets or ice crystals. The density of the accumulated water particles is also important, as is the color of the light incident upon and reflected by the clouds relative to the position of the observer and the light source. Clouds are described in terms of dimensions, that is, shape, diameter, height, structure, luminance, and color.

Clouds are of great interest to aviators as they are weather signposts in the sky (see Figure 4-1). They provide an indication of the vertical motions, winds, stability, and moisture in the atmosphere, and alert the smart pilot to potential weather hazards that may be lurking enroute. The clouds and their patterns will permit you to visualize the weather about which you have been briefed before takeoff. Each time a briefing is correlated with the clouds and weather encountered in flight, your aviation weather knowledge increases, and your next briefing will be more meaningful to you. You should recall that clouds undergo a continuous evolution into a variety of shapes. The international meteorological community has defined the characteristic forms most frequently observed and classified clouds into broad typical forms in terms of genera, species, and varieties.

An early classification of clouds was promulgated by Luke Howard in 1803, in an illustrated publication under the title, *On the Modification of Clouds.* Howard used Latin terminology to describe and name the clouds. These fundamental cloud types became the basis of the *International Cloud Atlas.* To be sure,

FIGURE 4-1 Clouds as signposts.

many others contributed to the system, including Renou, who in 1855 inserted middle clouds between the high clouds and low clouds in Howard's classification. Abercromby and Hildebrandsson published a classification of clouds which placed great importance to height as a criterion. Cumulonimbus clouds were described by Weilbach in 1880. The first *International Cloud Atlas* was published in 1896. The contribution of these observers and others were integrated to the classification until it evolved as it is presented in the *International Cloud Atlas* of today. The cloud-height categories generally cover a wide range of altitudes varying from sea level to 18 km (60,000 ft) in the tropics, 13 km (45,000 ft) in the midlatitudes, and 8 km (25,000 ft) in the polar regions of the earth.

The classification of clouds as given in the *International Cloud Atlas* is based on ten genera that are mutually exclusive: A cloud can belong to only one genus. By convention, that part of the atmosphere in which clouds are usually present has been divided to define three families of clouds: high, middle, and low. Each family is defined by a range of height at which the cloud genera most frequently occur. The high-level cloud family is composed of cirriform clouds including *cirrus* (Ci), *cirrocumulus* (Cc), and *cirrostratus* (Cs). The height of the bases of these clouds ranges from 3 to 8 km (10,000–25,000 ft) in the polar regions. In the warm temperate latitudes, the bases of high clouds are found between 5 and 13 km (16,500–45,000 ft). In the warmer tropical latitudes, the bases of these clouds range from 6 to 18 km (20,000–60,000 ft). The high clouds at these altitudes are composed almost entirely of ice crystals. The middle-level cloud family includes *altostratus** (As), *altocumulus* (Ac), and *nimbostratus*† (Ns). These clouds consist of primarily liquid water, much of it undercooled. The height of the bases ranges from 2 to 4 km (6,500–13,000 ft) at the poles; 2 to 7 km (6,500–23,000 ft) in the midlatitudes, and 2 to 8 km (6,500–25,000 ft) in the tropics. The low-level cloud family is made up of *stratus* (St), *stratocumulus* (Sc), and fair-weather *cumulus* (Cu) and *cumulonimbus*‡ (Cb). Low clouds are almost entirely water clouds, or undercooled water clouds in the cool seasons. These

---

*Altostratus is usually found among the middle clouds, but it often extends to higher levels.

†Nimbostratus almost always occurs as a middle cloud, but it can extend into other levels.

‡Cumulus and cumulonimbus clouds usually have their bases with the low clouds, but their vertical extent is so great that tops have been known to protrude into the tropopause and lower stratosphere.

clouds may also contain snow and/or ice particles when subfreezing temperatures are present. The height of the bases of low-level clouds ranges from the earth's surface to 2 km (6,500 ft) in the polar, temperate, and tropical latitudes. The following description of clouds is from the *International Cloud Atlas:**

cirrus (Ci) Detached clouds in the form of white, delicate filaments or white or mostly white patches or narrow bands. These clouds have a fibrous (hairlike) appearance, or a silky sheen, or both.

cirrocumulus (Cc) Thin, white patch, sheet, or layer of cloud without shading, composed of very small elements in the form of grains, ripples, etc., merged or separate, and more or less regularly arranged; most of the elements have an apparent width of less than 1°.

cirrostratus (Cs) Transparent, whitish cloud veil of fibrous (hairlike) or smooth appearance, totally or partly covering the sky, and generally producing halo phenomena.

altocumulus (Ac) White or gray, or both white and gray, patch, sheet, or layer of cloud, generally with shading composed of laminae, rounded masses, rolls, etc., which are sometimes partly fibrous or diffuse and which may or may not be merged; most of the regularly arranged small elements usually have an apparent width between 1 and 5°.

altostratus (As) Grayish or bluish cloud sheet or layer of striated, fibrous, or uniform appearance, totally or partly covering the sky, and having parts thin enough to reveal the sun at least vaguely, as through ground glass. Altostratus does not show halo phenomena.

nimbostratus (Ns) Gray cloud layer, often dark, the appearance of which is rendered diffuse by more or less continuously falling rain or snow, which in most cases reaches the ground. It is thick enough throughout to blot out the sun. Low, ragged clouds frequently occur below the layer, with which they may or may not merge.

stratocumulus (Sc) Gray or whitish, or both gray and whitish, patch, sheet, or layer of cloud which almost always has dark parts, composed of tessellations, rounded masses, rolls, etc., which are nonfibrous (except for virga) and which

*From the *International Cloud Atlas*, World Meteorological Organization, 1956. By permission of Secretary General, World Meteorological Organization.

may or may not be merged; most of the regularly arranged small elements have an apparent width of more than 5°.

stratus (St) Generally gray cloud layer with a fairly uniform base, which may give drizzle, ice prisms, or snow grains. When the sun is visible through the cloud, its outline is clearly discernible. Stratus does not produce halo phenomena except, possibly, at very low temperatures. Sometimes stratus appears in the form of ragged patches.

cumulus (Cu) Detached clouds, generally dense and with sharp outlines, developing vertically in the form of rising mounds, domes, or towers, of which the bulging upper part often resembles a cauliflower. The sunlit parts of these clouds are mostly brilliant white; their base is relatively dark and nearly horizontal. Sometimes cumulus is ragged.

cumulonimbus (Cb) Heavy and dense cloud, with a considerable vertical extent, in the form of a mountain or huge towers. At least part of its upper portion is usually smooth, or fibrous or striated, and nearly always flattened; this part often spreads out in the shape of an anvil or vast plume. Under the base of this cloud, which is often very dark, there are frequently low, ragged clouds either merged with it or not, and precipitation sometimes in the form of virga.

Since the advent of satellites, clouds are playing a much more important role in aviation meteorology. Aircraft are flying higher and faster, encountering more hazardous weather than ever before. All pilots should have a better understanding of the structure and lurking dangers of clouds that they encounter. The intimate characteristics of dangerous thunderstorms are often concealed in clouds of secondary importance. If pilots pay careful attention to the characteristic structure of these storms, they can identify and avoid the hazardous areas.

Clouds are the signposts in the sky which may be encountered along a route. Clouds, like fog, are often associated with poor visibilities and low ceilings, both of which are hazardous to safe flight. Clouds are at home in the sky, and pilots invade their environment. Air travel should generally be planned to avoid clouds, because clouds reduce visibility in all directions when you fly inside them. When you can't see, there is danger of colliding with other aircraft, or, if flying in mountainous regions, bumping into a granite-cored cloud. Clouds may also cause havoc with "dead-reckoning" navigation when you fly in or above them.

Clouds encountered enroute indicate to the experienced pilot weather conditions that can be expected in the immediate future. This information *may* correlate with cloud observations and weather information obtained at the pilot briefing prior to takeoff. (At least let's hope so!) This enables the pilot to check the timing of the forecasted movement of the weather systems discussed at the briefing, or to amend the forecast to agree more closely with the actual weather conditions encountered.

The degree of turbulence in the atmosphere can be ascertained by the cloud forms present. Cumuliform clouds are clouds of vertical development, which requires that the atmosphere be unstable. Thus, whenever cumuliform clouds are present, *watch out!* Expect to encounter rough flying conditions. By contrast, when stratiform or layer clouds are present, smooth flying is indicated. If you observe clouds with qualities of both cumulus and stratiform clouds, then you'd expect to find a slight degree of turbulence. In addition to being a good index of the degree of turbulence, cloud genera can also indicate precipitation. Clouds provide valuable information of moisture conditions in the local atmosphere. Since clouds form only when the atmosphere is either saturated or near saturation, an experienced aviator will know that immediately beneath the base of any cumulus cloud, very high relative humidities are present.

Clouds are usually telltale signs of flying weather. Every pilot should have an understanding of cloud genera, their development, organization, and relationship to flying weather. Complete and accurate cloud information will provide the pilot with enroute indications of the general meteorological conditions, and flying weather in particular, above the surface. Included in this information is evidence of atmospheric moisture content and stability. The cloud motions also give an indication of vertical air motion and the direction and speed of the horizontal movement (winds aloft). The continuity of the cloud array is important to all flight aspects, including navigation. The flyer always welcomes breaks in a cloud layer above or at the level of a planned flight. If these breaks are numerous, visual flight is possible.

An aviator is not only interested in cloud genera, but also in the base, height, and thickness of clouds along the route of flight. The temperature of the layer of atmosphere at the location of the cloud is also important knowledge to the pilot, as this suggests the possibility that structural icing will occur in the cloud. If clouds are used as signposts of the sky, the weather and its related hazards associated with each of the cloud genera should be understood. It is well known that high clouds (cirrus, cirrostratus, and cirrocumulus) do not usually affect flight operations, since they are usually thin and are composed of ice crystals, so structural icing is of no immediate concern. Cirrocumulus clouds are at times composed of small, undercooled water droplets that also cause no appreciable risk of airframe icing. High clouds, however, can be harbingers of future weather conditions that *may* cause air safety problems. Be alert!

Middle clouds of the altocumulus, altostratus, and nimbostratus variety present some hazards to aviation. The altocumulus (Ac) clouds are generally composed of water droplets. When undercooled, these clouds produce a kind of structural icing. Since these droplets are very small and very cold, they may not cause a serious airframe-icing hazard. Nevertheless, an icing danger does exist, and a smart pilot will avoid any conditions of possible icing. When flying below these clouds, virga may be encountered, and sometimes even rain and snow. Altocumulus clouds occur most commonly in extensive screen shroudings. Parallel streets of cloud rolls with fringed edges are often observed at multiple levels. Some surfaces may be jagged, with individual elements standing out in bold relief — exaggerated by natural shading.

Altostratus (As) clouds usually pose a greater hazard to aviation than altocumulus. They are grayish or bluish clouds striated in a fibrous or uniform appearance. They usually cover the entire sky, with some parts thin enough to reveal the sun above. Altostratus clouds, like altocumulus, are composed of water droplets, and ice crystals are present when the temperature is very cold. These clouds cover areas extending hundreds of kilometers horizontally, and of considerable extent vertically, as well. Altostratus clouds are quite dense even in the thinner portions and should be of great concern to pilots who may try to penetrate them. Overflying them poses no problem, but the thickness of these clouds will determine the amount of time a pilot must spend in climbing or letting down through them. The time factor is of major importance when icing conditions prevail in such clouds. The effective thickness of any cloud from the point of view of the pilot is a relative matter contingent on the time required to lift or let down an aircraft through a particular cloud.

Another problem that may be encountered when flying in or near altostratus clouds is precipitation. Often, this will be seen as virga trailing from a cloud base. When precipitation from altostratus clouds reaches the ground, the undersurface of the cloud layer acquires a ragged appearance. This type of precipitation is continuous rather than the intermittent type of cumulonimbus clouds. Altocumulus clouds should not be underflown, since both impaired visibility and structural icing may be encountered. An altostratus cloud layer often develops from a single layer of altocumulus. This occurs especially when ice crystal trails (virga) fall from the altocumulus.

Nimbostratus (Ns) clouds have a dark diffuse appearance with continuous precipitation, most of which reaches the ground. This type of cloud is thick enough to hide the sun. Beneath such a cloud, low, ragged cloud fragments are found which may or may not merge with the precipitating cloud. This cloud type also is composed of water droplets, which are often supercooled, and these droplets can cause lots of problems for the unwary aviator. Solid precipitation may also fall from these clouds as both snow and ice pellets (sleet). Nimbostratus clouds are very dangerous to fly over, under, or through—even in climb and descent altitudes. Often these clouds will break into layers while quickly merging, giving the pilot a momentary sense of false security. VFR flight underneath nimbostratus clouds is risky because of poor visibility and low ceilings along with freezing precipitation. Nimbostratus clouds are often confused with thick altostratus, but the nimbostratus type usually is a darker shade of gray and nearly always hides the sun and/or moon.

To conclude about middle clouds and the dangers they can present to pilots, we should remember that whenever altostratus clouds are thick and composed of undercooled water droplets, the threat of ice formation is always present. It is not prudent to penetrate such clouds, since icing conditions should *always* be considered potentially hazardous. A good rule of thumb is never to penetrate clouds unless it is *absolutely* unavoidable. Danger is always lurking nearby. Another rule of thumb is to avoid any cloud composed of water droplets when the free air temperature is below freezing. Middle clouds frequently produce precipitation and pose icing hazards to aircraft. Remember also that altocumulus clouds suggest some degree of turbulence in and adjacent to them, and this may serve as another harbinger of precipitation. To repeat, icing conditions are *always* present when undercooled cloud droplets are nearby. If a cloud composed of undercooled droplets is penetrated by an aircraft, structural icing will occur. Icing is also encountered when an aircraft flies under a middle-level cloud that is releasing liquid precipitation if the free air temperature is subfreezing. Underflying nimbostratus clouds nearly always brings the aviator and the aircraft in contact with precipitation, whether it be liquid, freezing, or solid.

Low-level clouds, like stratus or stratocumulus, or clouds

of vertical development, like cumulus or cumulonimbus, are clouds whose bases are below 2 km (6,500 ft). The stratus type forms generally as a gray cloud with a fairly uniform base, and is composed of small water droplets. At subfreezing temperatures, it may consist of undercooled water droplets or small ice particles. This cloud type most commonly occurs in a nebulous, gray, uniform layer. When it occurs at ground level, it is called fog. If its base is above the ground yet low enough to obscure the tops of low hills, high buildings, or towers, it is called stratus. Its undersurface, when above the ground, is usually well defined and may show undulations. If it does form on the ground, it may be lifted by surface warming to a few hundred meters above the surface. Flight beneath these low-level clouds is not recommended, since clearance between the cloud and ground is limited, and danger exists that such a flight would encounter a topographic or anthropogenic projection, such as a knoll, hill, building, or tower. Visual flight below stratus clouds is often impossible. The clouds are frequently so dense that a view of the ground is difficult to maintain, and this problem is particularly hazardous during descent or landing. Stratus clouds are easily identified because of their absence of any form. Precipitation isn't usually associated with stratus clouds, but when present it is either drizzle or very light rain. Turbulence is rarely found in stratus clouds.

The stratocumulus is a whitish-gray layer cloud composed of rounded masses and rolls that are nonfibrous and are regularly arranged. This type of cloud is composed of water droplets, and sometimes raindrops or snow pellets. It most often appears as a layer of cloudlets similar to altocumulus but at a lower altitude. Sometimes the elements are in the form of parallel rolls separated by cloudless streets. These clouds often appear to be resting on several levels simultaneously. Flight in and near stratocumulus clouds may encounter slight to light turbulence. Any flight in low-level clouds (as in all other clouds) poses the problem of poor visibility. Unless you are IFR-rated and proficient, stay away from these clouds.

Stratus fractus is another low-cloud species, with irregular shreds, ragged in appearance. Other low-level clouds are those of vertical development, the cumulus and cumulonimbus. The cumulus cloud develops vertically in the form of rising domes or mounds. These are rather easy to avoid, and they possess at least light to moderate turbulence. The cumulonimbus cloud is also known as the "mother of clouds," because after they lose their energy and are in the dissipating state, they break down into cirriform clouds, altoform clouds, and stratocumulus. These clouds should be avoided *at all costs!* Such clouds have bases ranging from 1,000 ft in the flatlands to 10,000 ft in the mountains and their tops have been reported to exceed 70,000 ft at times. Thus, overflying them is not a practicable idea. Clouds of this type are some of the most dangerous an aircraft can encounter, no matter how briefly. The cumulonimbus clouds have severe to extreme turbulence, thunder, lightning, heavy rains, and showers with hail, gusty winds, and icing. They are very dangerous even to think about flying in, since their vertical currents buffet a plane into many configurations of attitude. There is no worse enemy an aircraft, or anything airborne, can become involved with. Often even large aircraft that encounter cumulonimbus clouds and their extreme turbulence sustain structural damage of great severity.

Flight plans must be made to conform with existing and forecast weather conditions. Smart VFR pilots always plan their flights so that they can maintain visual contact with the ground at all times. The presence of a cloud layer will often prevent the pilot from taking advantage of the most favorable winds. The aviator should remember that clouds are associated with many aviation weather hazards (icing, turbulence, poor visibilities, low ceilings, visual obstructions, and granite-cored clouds, to list a few). When flying, always use clouds as signposts.

5 AIR MASSES, FRONTS, LOWS, AND HIGHS

AIR MASSES   What is an air mass? It is a body of air which is essentially homogeneous in its internal distribution of temperature and humidity and which also is separated from an adjacent air mass by a transition zone (*front*) in which these qualities vary, sometimes a great deal. The life history of an air mass may be divided into three distinct parts. Beginning with its source region, an air mass acquires its basic temperature and moisture properties. Later, the properties of the surface over which it travels influence the air mass significantly; terrain is nature's modifier. Finally, the air mass dissipates its original characteristics, as the air mass moves still further from its source.

Air masses are generated whenever a body of air approaches and remains over an air-mass source region. Imagine a surface that is warmer than the air over it, causing the air to be heated from below. The heat of the surface warms the air as the surface heat is shifted to the atmosphere by conduction, convection, and radiation (see Figure 5-1). The main amount of heat shift would be caused by radiation, although convection and conduction contribute to this exchange. The conduction of heat between a solid and gaseous medium (at the interface) is small, so convective heat transfer is limited in a stable air mass. Thus, a body of warm air over a cool surface would assume the temperature and moisture characteristics of the surface. The net exchange

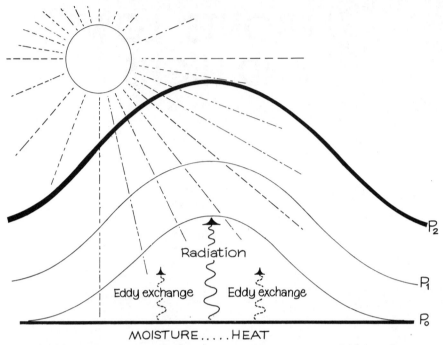

FIGURE 5-1 An air mass assuming the moisture-heat characteristics of the air mass source region.

would be from the surface. If the surface were an ocean, sea, or moist land surface, the body of overlying air would gain moisture even if it remained over the surface for only a few days.

A valuable property of the air-mass concept is that the air mass retains its identity for several days after it moves from the source region. The primary source regions feature permanent and semipermanent anticyclones (high-pressure areas), which are prominent within the planetary general circulation regime (see Figure 3-2). For example, maritime source regions in the winter are the northern Pacific, the Gulf of Alaska, and the northern Atlantic. These areas are the home of maritime-polar air masses that affect North America. A continental-polar air-mass source region is found in central Canada and the north central United States. This source region has an effective wintertime snow cover. Snow cover, to be effective, must have a -3°C (27°F) mean isotherm during the coldest month. If the isotherm (line of equal temperature) is only 0°C, the snow cover will be "spotty," and will barely occupy the perimeter of the continental source region.

Air masses are classified according to their source regions.

Some of the characteristics of the Canadian source region have already been described. There are numerous other source regions on the earth's surface with a quasi-homogeneous surface structure, over which air may remain long enough to arrive at an equilibrium with the temperature and moisture properties of the surface. Air-mass source regions are found at locations on the earth's surface where anticyclones dominate the pressure field. One such case occurs in central Canada during the winter months, as was previously cited. Air masses can also be classified by latitude. Thus, the polar air masses are cooler than the subtropical air masses. Latitudinal classification defines four air mass source regions:

1. Arctic (A) — an extremely cold air mass of the polar regions, generally originating in Siberia and northern Canada.
2. Polar (P) — a cold air mass of the polar regions.
3. Tropical (T) — a warm tropical air mass of the subtropical high.
4. Equatorial (E) — a warm air mass originating in the equatorial convergence zone.

The difference between the arctic (A) and polar (P) air masses is small, specifically, the severity of coldness. The polar and tropical air masses are very important in midlatitude meteorology and especially weather forecasting. These air masses are again subdivided, to better describe them, according to whether their source is over a land or water surface. Those forming over a land surface are called *continental* (c), and those forming over a water surface are called *maritime* (m). The underlying surface plays a significant role in determining the characteristic temperature and moisture properties of air masses.

Source regions of polar air masses are found in the polar highs. The tropical-maritime air masses of interest to aviators and meteorologists in the Northern Hemisphere are found near Bermuda or the Azores, or placidly floating over the Gulf of Mexico. In the tropical Pacific, maritime air masses extend from the west coast of Mexico and the United States to the international date line! Hawaii may be such a center. The continental-tropical air masses are found near the same latitudes as is maritime tropical air, but over the continents instead. In North America, this type of air mass is found in summertime over deserts of the southwestern United States and northern Mexico. These are also found in northern Africa and southwestern Asia.

Do various air masses have a thermodynamic character, as they move from their source region to newer surfaces? Will the air mass be colder or warmer than the new surface over which it arrives? Talphas Bergeron, a Scandinavian meteorologist, suggested that air-mass classifications should include a term that would better reflect its thermodynamic character, so he introduced yet another classification: (1) k (cold), air masses that are colder than the underlying surfaces; (2) w (warm), air masses that are warmer than the underlying surfaces.

Aren't all these classification systems confusing? Yes, they are, but the last one has utility for pilots. A combined classification system is shown in Table 5-1. The first letter of the classification, given in lower case, defines the surface over which the air mass originates, such as maritime (m) or continental (c). The second letter, in capitals, denotes the latitude (temperature) of the source region, such as tropical (T) or polar (P). The third letter, again lower case, denotes the thermodynamic classification of Bergeron, that is, cold (k) or warm (w).

To a pilot, the third letter would indicate thermodynamic character — or relative stability — of the air mass. The w air mass (which is warmer than the surface over which it resides) suggests stable air, which indicates to the aviator that flying should generally be smooth. Using the same reasoning, the k air mass (which is cooler than the surface) suggests unstable air, which means a possibly bumpy flight. We can go into more detail along these lines. A warm (w) air mass might display the following:

1. Smooth flying conditions.
2. Stable lapse rate.
3. Poor visibility near urban and industrial areas.
4. Stratiform clouds and fog.
5. Drizzle.

A cold (k) air mass might display the following:

1. Turbulence from surface to 10,000 ft.
2. Unstable lapse rate.
3. Good visibility except in precipitation.
4. Cumuliform clouds.
5. Showers.

TABLE 5-1 A General Air-Mass Classification

Surface of Origin	Latitude	Thermodynamic Classification	Symbol
continental	Polar	cold	cPk
continental	Polar	warm	cPw
continental	Tropical	cold	cTk
continental	Tropical	warm	cTw
maritime	Polar	cold	mPk
maritime	Polar	warm	mPw
maritime	Tropical	cold	mTk
maritime	Tropical	warm	mTw

FRONTS The concept of fronts has only been understood and used since 1918, when a group of Scandinavian meteorologists proposed the "polar front," the semi-permanent, semi-continuous boundary which divides polar and tropical air masses and on which most of the storms of the midlatitudes form and move. The polar air-mass boundaries have a seasonal undulation, advancing toward the equator during the spring and early summer and then retreating toward the poles during the fall and early winter. During the summer months, the polar frontal surface is less well defined.

History has taught us that weather, especially weather affected by the polar front, has had a great influence on the outcome of the battles of war, particularly those fought at sea. Much to their chagrin, the British and French discovered the hazards of weather during the Crimean War; the loss of more than two dozen ships in a vicious storm on the Black Sea heavily contributed to the defeat of their fighting forces. The British rejoiced over the fact that weather played a major role in the defeat of the Spanish Armada in 1515. It was in light of these circumstances that the Allies and Central Powers each tried to deprive their enemy of valuable weather information during World War I. This wartime curtailment left the Scandinavian maritime fleets without needed weather information for their own safe and successful operations. Norway's meteorologists were challenged to provide better weather forecasts using limited weather information (especially in the northern Atlantic). In the research that ensued, they discovered the polar-front and air-mass concepts of weather analysis, a tremendous leap for the science of meteorology. They discovered a relatively narrow transition zone (boundary) between the warmer tropical air

62 *Air Masses, Fronts, Lows, and Highs*

and the colder air from the polar regions. This transition zone was called a *front*, a word that was in vogue with the newspapers of that era, formerly used to describe the lines of battle between armies. In meteorology, a front may be thought of as a battleground between conflicting air masses.

As the air mass that engulfs the polar area was studied, it was discovered that this mass of cold air, with its undulating perimeter, was difficult to understand. These undulations sometimes protrude toward the equator. When this happens, a frontal system is born. Between the migratory polar air and the warm

FIGURE 5-2 Cold front.

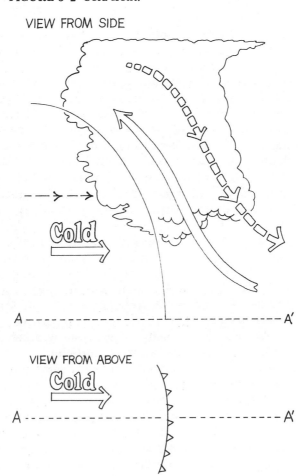

tropical air is a frontal surface. This surface is a sloping zone of transition separating two air masses, each having its own density, temperature, and humidity structure. The transition zone, or front, is represented on a weather chart as a line along which the frontal surface intersects the surface of the earth. This line represents an air-mass border region, which is usually quite narrow (10–50 km). Frontal surfaces have very gentle slopes varying from 1:50 to 1:300 (1 km on the vertical axis to 300 km on the horizontal). Converging and rising air are important features of well-defined fronts. Upward motion results from the mechanical lifting of the warm air as advancing cold air slides underneath. The rising and adiabatic cooling of the warm air result in condensation, clouds, and precipitation, which mark the typical active front. Measurements in the free atmosphere show that horizontal gradients of temperature within the separate air masses are not negligible. Fronts generally lie in troughs of low pressure often found between air masses. The more extensive fronts, such as polar fronts, sometimes extend from the surface to the tropopause. Principal frontal types include cold, warm, stationary, and occluded varieties. The leading edge of an advancing cold air mass is defined as a cold front. At the surface, the advancing cold air overtakes and displaces (slides under) the warmer air, which then glides overhead. Cold fronts, depending on the wind speeds behind them, often travel at speeds of about 25 to 30 knots. Figure 5–2 shows the vertical cross-section of a cold frontal system and the symbols representing it on the surface weather chart.

    The warm front differs from the cold front in that warmer air advances upon and eventually overtakes the cold air mass it will replace at the surface. Since the warm air is less dense than the cold air, it rises in a gentle gliding motion over the denser cold air. Cold air retreats, while hugging the ground. This retreating process of the cold air produces a more gradual frontal slope than that of the cold front. Consequently, the warm front on the surface is seldom as well marked as the cold front (see Figure 5–3). Warm fronts move at about half the speed of cold fronts, since the wind flow is generally from the south around a low-pressure area. Stationary fronts do not move (see Figure 5–4). The opposing forces mobilized by the adjacent air masses of different densities are such that a stagnant frontal system, which separates the air masses, shows little movement. In such cases, the surface winds tend to blow parallel to the frontal surface. Although the slope may be steep, depending on such factors as wind distribution, density differences, etc., the usual slope of a stationary front is shallow, because the frontal systems are in active opposition. Frontal waves, frontogenesis, and cyclogenesis often develop along stationary fronts.

The upper wind flow dictates to a great extent a cyclone's life cycle. Figure 5-4 shows the initial conditions of frontal development; the wind flow is parallel to the stationary front. A small disturbance may initiate a cyclonic deflection of the flow pattern, which in turn may form a frontal wave. If this perturbation in the circulation persists, the frontal wave usually increases in intensity. If the eastward section of the front begins to move poleward as a warm front, then the western section of the former stationary front moves southeastward as a cold front. This deformation initiates both the cyclonic and frontal-type wave. Because of the cyclonic circulation, the pressure at the center of the frontal wave may fall; if so, a closed low-pressure area forms

FIGURE 5-3 Warm front.

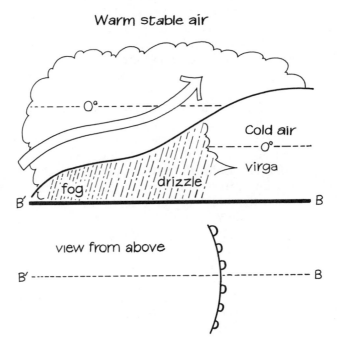

near the center. As cyclonic circulation increases, the surface winds become strong enough to cause the frontal system to begin to move. The cold front moves much faster than does the warm. Because of the difference in the speeds of the cold and warm fronts, the cold front overtakes the warm, and the warm front finally obstructs the movement of the cold front, resulting in an occluded front. At this juncture, the cyclone becomes mature and attains maximum intensity. The symbols depicting the occluded front are a combination of those used for a warm front and a cold front. As the occlusion continues to enlarge, the cyclonic circulation diminishes in intensity, and the frontal movement slows. Sometimes a new wave develops on the trailing part of the front, or a secondary low-pressure area forms where the warm and cold fronts form the occlusion. Eventually, the two fronts in the occlusion merge into a stationary front after the energy is dissipated.

Figure 5-5 shows the warm-type occlusion in a vertical cross-section. This occurs when the cool air ahead of the advancing warm air is colder than the cold air of the cold front. The cold air of the cold front overrides the colder air in advance of the

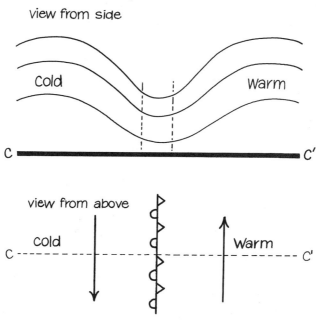

FIGURE 5-4 Stationary front.

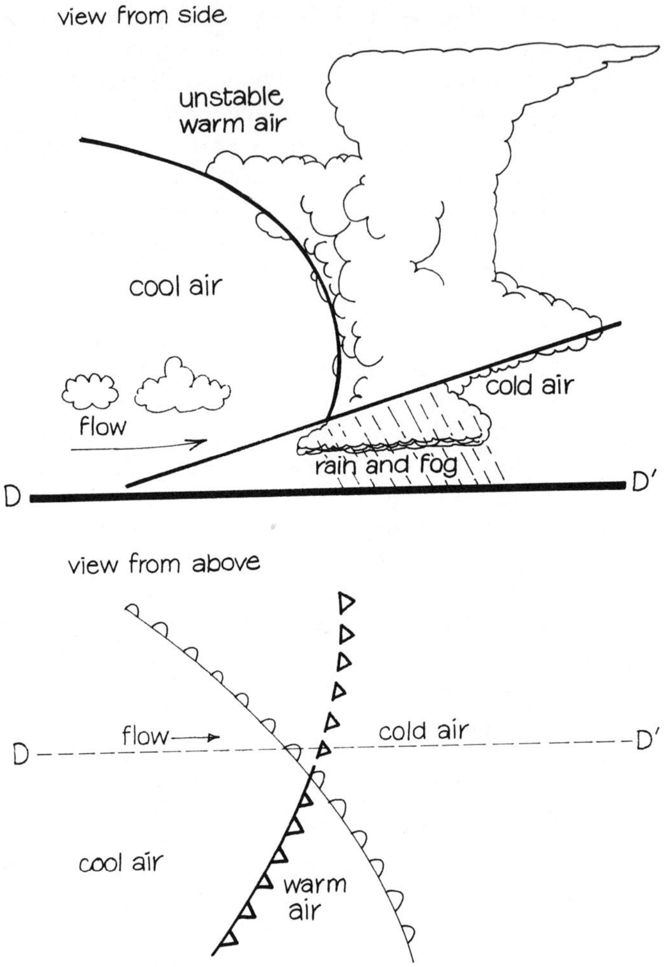

FIGURE 5-5 Warm front occlusion.

warm front, and the cold front with its cool air is lifted as it passes above the cold air preceding the warm front. The warm frontal air is elevated above the cold front, so we find three distinct layers of air laying horizontally above the surface—forming a warm-type occlusion.

A cold-type occlusion is shown in Figure 5-6. This type of occlusion occurs when the air following the cold front is colder than the air in advance of the warm front. Like the warm occlusion, this creation has three distinct layers of air. It should be noted in all cases that cloud types associated with the various

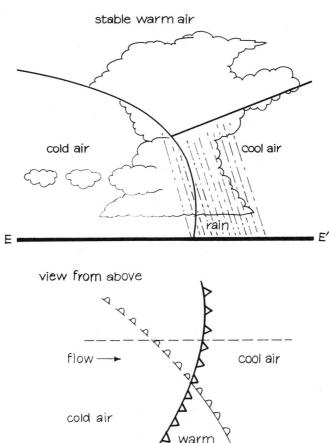

FIGURE 5-6 Cold front occlusion.

fronts and occlusions will be governed essentially by the moisture and stability of the *warm* air in the system.

Low-pressure systems are not always accompanied by fronts. Often, nonfrontal lows form in mountainous regions of the western United States, especially on the lee side of mountains where a closed low or lee-side trough often develops. Examples of such systems are the *Colorado low*, which forms on the east slope of the Rockies. If this type of system forms on the lee side of the Sierra Nevada Mountains, it's referred to as the *Nevada low*. Still another type of low-pressure system is formed in the southern Arizona and northern Mexico deserts of North America, and is called a *thermal low*. This low-pressure system is rather shallow; its relative strength decreases rapidly aloft, where it becomes a *high-pressure* system (a warm-core low). This low does not have fronts associated with it, either. Tropical lows also are not associated with frontal systems.

We are also interested in the life cycles of frontal systems. This includes the formation (*frontogenesis*) and the decay (*frontolysis*) of fronts. Fronts just don't suddenly spring into being; they appear only after frontogenesis has occurred for some time. Similarly, fronts don't suddenly vanish from the surface; they disappear only after the process of frontolysis has been working on them for a period of time. Frontogenesis is a Latin-derived word meaning front-creating. Frontolysis is a hybrid Latin and Greek word meaning front-dissolving. Frontogenesis occurs when the wind field converges. This forces the isotherms (lines of constant temperature) to move closer together. Either the temperature field or wind field must become nonuniform for the isotherms to become concentrated. If the two fields are uniform, the isotherms will move without any concentration of spacing.

Frontolysis occurs whenever the wind field is divergent and the isotherms become spread out. As such, divergence of the wind field actuates frontolysis. Frontogenesis occurs with a concentration of the temperature field (isotherms moving closer together), and is especially likely near the borders of an air mass, or along the shorelines of continents.

Let's consider the weather sequence that could be expected to occur at each of the frontal systems under idealized circumstances. We must understand that weather disturbances are usually very complex, and that weather systems in the real world differ greatly from idealized models. But many weather agitations contain elements of similarity which correlate very well with conditions of the theoretical models, so knowledge of these similarities can be helpful in better evaluating the real weather phenomena.

Warm fronts represent the replacement of cold air by warm air. The warm air forms a shallow frontal surface above the slow-sliding air of the retreating polar air mass. We can reason that because of the slow rate of rise, vertical lifting of the warm air will not be as great, particularly when the warm overriding air is stable. This phenomenon resembles a large-scale lifting of air. If the overrunning air is stable, clouds will usually be stratus types, forming high above the freezing level. Cirrus clouds will act as harbingers of the approaching warm front; they may form several hundred kilometers in advance of the frontal position. The presence of these and other clouds encountered in flight provides an experienced pilot with the weather signposts of the sky. By understanding the characteristics of various cloud genera, an aviator can correlate the weather briefing with the actual weather! This enables a pilot to immediately amend any briefing forecast received prior to takeoff to agree with the actual weather conditions that are being encountered. Watch the sky's signposts! Even if you are not flying, you'll know how to dress for tomorrow's weather. The basic information obtained from the clouds and the patterns of their arrays assists the flyer to better comprehend and forecast the flying weather associated with them.

A general rule of thumb may be used to predict turbulence: Cumulus-type clouds are the precursors of rough air. The greater the vertical development, the more severe the turbulence. Stratus-type (layer) clouds are evidence of a less turbulent or smoother flight. Clouds possessing characteristics of both (stratocumulus, altocumulus, and cirrocumulus) suggest less turbulence than would be found in clouds of great vertical development. The severity of the turbulence is significant if the flight penetrates the cloud, but these clouds can also forewarn a pilot of which type of precipitation can be expected, if any. These include showers (large drops), rain (intermediate-size drops), and drizzle (small drops). Shower precipitation is from cumulus and cumulonimbus clouds; steady rain and/or snow are phenomena of the nimbostratus and altocumulus clouds. Drizzle is associated with warm moist air rising over colder and dense air at the surface, as is found in a warm front. Clouds may also provide a useful index of the moisture content of the atmosphere, since high relative humidities are found near their bases. Knowing this can assist you in avoiding problems of power-plant icing.

Let us examine clouds as they relate to flying. Clouds exert great influence on flight. Any logical person is aware that concealment in clouds constitutes a hazard that increases the risk of collision with other aircraft and anything else inside a cloud (like the top of a mountain). This "riding within clouds" can also create dead-reckoning navigation difficulties and at times bring about icing conditions. To be specific, just below the bases of cumulus clouds the relative humidity is particularly high. Also, when lenticular clouds form over ridges or mountain tops, high relative humidities are always present. A pilot flying under these conditions should always anticipate carburetor icing.

While the implications of cumulonimbus clouds are easily

understood by the majority of pilots, the weather implication of other frontal clouds are less likely to be perceived, even when a pilot is familiar with the implications of frontal weather. Here is how to notice the high- and mid-level clouds found along and near your flight path and to appreciate the useful cloud indicators that are available. Visualize the cloud structure of a warm frontal system approached from the cold air (as might be encountered while flying from north to south in midcontinental North America). The first sign that we are approaching a frontal cyclonic system is the appearance of high clouds. We perceive these to be cirrus changing to cirrostratus while continuing to form a denser cloud sheet covering the celestial dome. Now, if the warm air were unstable, cirrocumulus would hover into sight. (The sailors of old would call this a "mackerel sky.") As we fly near the warm front, we encounter clouds of increasing density; now, instead of having high cirrus and cirrostratus clouds overhead at about 6,500 m (20,000 ft), we are surrounded by mid-level clouds of the altostratus and altocumulus varieties. As these clouds thicken, we encounter virga. The cloud array continues to thicken and lower; as we approach the stratus, stratocumulus, and nimbostratus clouds, precipitation of either rain or snow (depending on the season) is encountered. We could try descending to underfly the system or perhaps find flying room between cloud layers. But if the warm air is unstable, an embedded cumulonimbus might be hidden from our view. If we are very lucky we might avoid the cumulonimbus; on the other hand, if we do confront one our demise may be reported by the local media! If our flight misses the worst dangers and hazards and we are still airborne, we will be jostled as though we were riding a bucking bronco! Rain or snow usually falls from the cloud deck overhead. If the freezing level is low enough, the precipitation will be of the freezing variety, which can "down" the aircraft in minutes. Finally, the frontal zone is passed, and we approach the outside boundary of the frontal system. At this point in time, we realize that we have been through a hair-raising experience, inasmuch as that imbedded cumulonimbus provided its usual thunder, turbulence, and lightning pyrotechnics, which did anything except calm our shattered nerves! So much for flying under the system!

In a flight on top of a warm front, the frontal structure will not be as obvious as in a mid-level flight. If thunderstorms are embedded in the warm frontal array of clouds, they can easily be seen and avoided "on top," and the front's position will be clearly shown by the ending of the cloud deck below. Thick nimbostratus clouds soon change to warm-sector stratus or to a cloudless sky. During the summertime, showers and thunderstorms will be observed in the warm sector when the air is unstable and sufficient moisture is available. These air-mass clouds differ considerably from the frontal clouds that have been described.

If we fly away from a warm front in the cold air, the sequence of clouds will appear in reverse order. We quickly note as we depart the frontal activity that our flying weather is rapidly improving. The lower clouds vanish as the bases of the middle clouds slowly rise. As the cold air is penetrated the cloud genera change from nimbostratus to the less dense altostratus. Precipitation becomes intermittent, evolving to virga, which finally ceases. (But if the temperature of the cold-air mass were to be below 0°C, the liquid precipitation would freeze on contact with the aircraft. We'd then experience heavy icing conditions, which could be fatal — or at least chilling! Please excuse the pun!) As the altostratus clouds become less dense and sparsely distributed, the clouds change to the cirriform type (cirrus, cirrocumulus, and cirrostratus), and these will evaporate in the cool dry air, leaving a cloudless sky.

On those occasions when a trough of low pressure is present without a discernible surface front, the array of clouds, weather, and general flying conditions closely resemble those of a warm front, just like those situations when warm moist air overrides cool dense air.

If our flight approaches an active cold front from the warm sector of the cyclone, a large mass of cloud of considerable extent will be seen. If the warm air is moisture laden and unstable, the clouds will be cumuliform. Remember, cumulonimbus clouds demand respect since they contain a great many of the common hazards to aviation. These giant cumulonimbus tower upwards of 15 km (50,000 ft); their debris of anvil cirrus stretch out in great sheets a hundred or so kilometers downwind, signaling the approach of the frontal system. After the cirrus, a well-organized layer of altocumulus clouds will be found. The closer the frontal surface is approached, the thicker the cirrus clouds become.

Such a cold front will appear slightly different when viewed from the cold sector, since the overhead cirrus may not be present. A huge mass of cloud will reach from 500 m to heights greater than 15,000 m, and its massive proportions will literally overwhelm any unwary invader. If you love life — and want to keep your aircraft intact, don't fly through cold fronts! Severe to extreme turbulence is but one of the dangers to be encountered. Another is precipitation, ranging in size from intense showers to tiny floating droplets of drizzle. When the air is unstable, cumulonimbus clouds become thunderstorms and rise to great heights. Care to confront a belching, spitting, sparking cumulonimbus cloud? You'll be sorry! These clouds — which build in unstable, warm, moist air — often develop into an organized system of cumulonimbus clouds. Here is born the most violent weather of all: heavy precipitation, high winds, hail, severe to extreme turbulence, icing, and even tornadoes. If you are piloting an airplane and approach a storm of this description, you would do well to make an 180° turn and look for an airfield; then land as quickly as possible, tie down the machine securely, and let the storm have sole possession of the heavens. Such storms have priority use of airspace over and beyond the demands of any humans. After all, even commercial airliners avoid a direct confrontation with such storms whenever possible; so what chance do you think a light plane has?

A cold frontal system frequently triggers thunderstorms at its surface and also a hundred or so kilometers ahead in warm, moist, unstable air. These are called instability lines, also known as squall lines, and they are often dotted with severe thunderstorms. These nonfrontal thunderstorms also are not safe to be near in a flying machine. Squall-line thunderstorms are very similar to frontal thunderstorms; both are *very* dangerous to aircraft operations. They frequently spawn severe weather conditions, like high surface winds, large hail pellets, and even tornadoes.

An occlusion presents a cloud system different from that of either a cold front or a warm front; it has a cloud array of its very own. In a flight through such a system, the pilot will observe cirrus clouds increasing and thickening into cirrostratus clouds, which lower into altostratus. Precipitation may be present, frequently forming low scud clouds. Nimbostratus clouds are encountered next, with precipitation increasing in intensity and cloud bases lowering; in many cases, cumulonimbus clouds are concealed in the nimbostratus. This is always a chance someone takes when flying near or in an occlusion at low or middle levels of flight. The cloud mass ends abruptly on the west side of the system. A frontal occlusion will present sheets of altocumulus clouds; both the warm- and cold-type occlusions have such an appearance from the air. The extent of vertical development of the clouds depends very largely on the stability and moisture content of the air masses, and the speed of movement of the system. A flight in the opposite direction through the occlusion will reveal the same array of clouds, but in reverse order.

Flight plans must be made to conform with existing and forecast weather conditions. A VFR pilot must always plan flights so that visual reference with the ground is uninterrupted at all times. The presence of a cloud layer will often prevent pilots from taking advantage of the most favorable winds.

Surface-weather charts show fronts and other hazards of flight, and other forms of chart analysis allow pilots to determine the expected weather conditions along the route of flight. Knowing the location of weather systems will assist in planning a safe air adventure! The pilot is, naturally, the best observer of current weather conditions while in flight. This is because weather stations are spaced from 85 km (50 miles) to 170 km (100 miles) apart at the surface, and the upper-air stations are spaced from 250 km (150 miles) to 500 km (300 miles) apart. Thus, they do not always sufficiently describe the actual flying weather that will be encountered. Remember, too, that frontal weather changes rapidly unless you are caught in it. Alertness is required to evaluate each situation as to the appropriate course of action. It is an excellent idea when planning a flight to have an understanding of weather systems enroute and then plan an alternate route to take to avoid any hazardous situation that may confront you while in flight. This is important because you will be able to avoid making an emotional decision while directly involved with the weather. The decision will have been made on the ground, calmly.

During your preflight planning of alternate routes, remember your limitations as well as those of your aircraft. A major part of understanding weather conditions is knowing enough to avoid those situations with which you have little chance to cope safely. This requires that you be honest with yourself about what you *can* and *cannot* do in a given aircraft. Don't make any news today that you will not be around to read or hear about tomorrow!

6 THUNDERSTORMS

Most of us at one time or other have witnessed an awesome display of lightning and thunder. Thunderstorms may usher in hail, torrential rains, strong gusty winds, and once in a while the most violent and intense of all local effects—tornadoes! Thunderstorms are packaged in many shapes, sizes, and intensities. They vary from an occasional flash of lightning with its accompanying dull rumbling of thunder to the severe local storm. The sheer power of these horrendous natural displays from the massive cumulonimbus clouds can literally overwhelm you. Thunderstorms are common and familiar. They often announce their coming by the debris of the anvil cirrus stretched out in great sheets a few hundred kilometers in advance of the giant cumulonimbus clouds of the storm system. This telltale sign should give the weather-watching pilot sufficient time to cope with this weather situation—by getting the *hell* away from it! Too many pilots have the Scarlet O'Hara syndrome: "I'll think about that tomorrow." To a pilot, that can be fatal.

By agreement in the World Meteorological Organization, a thunderstorm is reported at a station when thunder is heard, or when overhead lightning or hail is observed and the local noise level is such as might prevent hearing thunder. Thunder is the audible noise of lightning discharge. Thus, thunderstorms are defined in terms of their electrical manifestations. It is true that a thunderstorm is a local storm spawned by a cumulonimbus cloud, always accompanied by lightning and thunder, usually of short duration, seldom lasting over 2 hours for any one storm. To the meteorologist, the thunderstorm is the consequence of atmospheric instability and constitutes, loosely, an overturning of air layers in order to achieve a more stable density stratification. In order to obtain a more complete understanding of a thunderstorm's birth, growth, and development, we should begin with a discussion of thermal convection. As used in meteorology, this refers to atmospheric motions that are predominantly vertical, resulting in the transport and mixing of atmospheric properties. We must differentiate convection from advection, which refers to horizontal transport and mixing. Convection is initiated by air that is lifted or raised, so a lifting mechanism is required.

To get a better understanding of the convective process, let's indulge in another simple mental experiment: First, visualize the atmosphere as being composed of many small bubbles of air, *each* encased in a weightless, elastic cover. This cover does impede the exchange of heat with the environment. Convection commences as each small bubble of air on or near the ground is warmed at the surface by conduction and radiation. When air bubbles are heated, their molecular activity increases, causing their volumes to expand, with a corresponding reduction in density. Then, buoyant forces initiate vertical motion. As an air bubble rises, it cools at the rate of 1°C/100 m (5.5°F/1,000 ft); this is the *dry-adiabatic lapse rate*. An air bubble is buoyant as long as its temperature exceeds that of the environment.*

If sufficient moisture is present in the imaginary small bubble, and it continues its upward motion, then its relative humidity will soon approach saturation (100 percent). The ability of the atmosphere to retain water vapor is a direct function of its temperature. At this point in its upward journey, condensation will commence. Now, a cloud becomes visible (from all the many small bubbles that have risen), and the adiabatic cooling changes from the dry to the moist lapse rate. This means that rising saturated air, instead of cooling at a rate of 1°C/100 m, cools at the rate of 0.5°C/100 m. Because the moist adiabatic cooling is at a reduced rate, the former conditionally stable air becomes unstable. The condensation process releases kinetic energy in the

---

*At any level in the atmosphere, a vertically displaced air bubble will cool at the dry adiabatic lapse rate. If the environmental lapse rate is greater than the dry adiabatic lapse rate, its vertical motion will be quickened. But if the environmental lapse rate is less than dry adiabatic, vertical motion will be inhibited.

form of vertical motion to the atmospheric layer where clouds are forming. It should be recalled that for each gram of water vapor condensed into the liquid state, approximately 600 calories are released to heat the air. Thus, the air becomes more buoyant, and vertical motion is generated.

Now would be a good time to consider the contribution water vapor makes to the buoyancy of air. This contribution is due to the difference in molecular weights of water and dry air. Water vapor is a lighter gas than dry air: The molecular weight of water vapor is 18; that of dry air is 29. It may sound logical to say that when a volume of water vapor is added to dry air the total weight will be 47. Alas, this is not the way gases are combined! Let us combine dry air and water vapor so that the resulting volume will be 96 percent dry air and 4 percent water vapor. Then air's true molecular weight is found as follows:

$$0.96 \times 29 + 0.04 \times 18 = 28.56$$

Obviously, the more water vapor in the bubble and in the environment, the lighter the weight of the air and the greater the buoyancy. Each of these make a substantial contribution in determining atmospheric stability.

Since we are discussing vertical motions, we see that our bubble of air will accelerate when its temperature is warmer than the temperature of the environment. For the atmosphere to be unstable, it is not enough to have the temperature decrease with altitude. If this were the only reason for instability, then we would experience a stable atmosphere only when the temperature increased with altitude. This sort of temperature distribution is called an inversion (the opposite of its normal distribution). Inversions are not uncommon; they occur whenever the ground cools more rapidly than the overlying air.

In order for the atmosphere to be unstable, the rising air bubble must always be warmer than its environment. When the air rises, it cools at the dry adiabatic lapse rate until saturation begins. For absolute instability (see Figure 2-2), the lapse rate of the natural environment must be greater than dry adiabatic; when this condition is present, the atmosphere is said to be absolutely unstable. Environmental lapse rates of these magnitudes are found mostly over the hot, dry desert regions of the world, in the layer of air immediately adjacent to the surface.

We have observed that a rising bubble of air is cooled by adiabatic expansion. As the rising air is cooled, the moisture content does not change until condensation begins. Thus, as the air parcel cools, the relative humidity increases. Once condensation begins, the conditions of instability become less restrictive. An environmental lapse rate that lies between the moist adiabatic lapse rate and the dry adiabatic lapse rate is said to be conditionally unstable. This means that the bubble of air is stable when compared with the dry adiabatic rate but unstable when likened to the moist adiabatic rate.

Now, let us see how this is to be managed. First we take a bubble of air and lift it adiabatically while it cools at the dry rate. We must keep in mind, however, that if it contains moisture, it may become saturated after rising a few thousand feet. Thus, condensation commences, and a cloud becomes visible. Once the water vapor condenses, latent heat is released. Once saturation has begun, the imaginary bubble of rising air cools at the moist rate. The released latent heat finally warms the bubble. As it becomes more buoyant, it travels to a higher level with its new energy supply.

This is the first step in the making of a thunderstorm! When many of these bubbles rise in clusters, a cumulonimbus cloud is created (see Figure 6-1). These clouds form the cells of a thunderstorm. The cumulus stage is a thunderstorm in its full glory. In the next step, the cloud continues to grow until precipitation begins to fall. In the mature stage, rain or snow causes the cloud to develop downdrafts. Suddenly, the thunderstorm

FIGURE 6-1 The three stages of a thunderstorm.

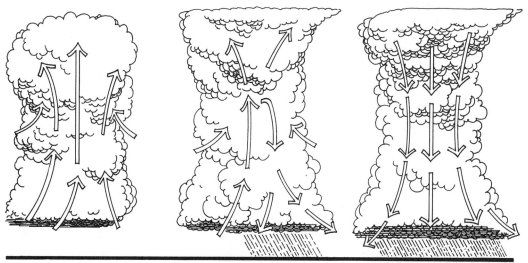

The Cumulus Stage   The Mature Stage   The Dissipating Stage

becomes suicidal and starts to destroy itself. As time passes, more updrafts are converted by the precipitation into downdrafts. At last, the entire cloud has turned into one large downdraft! This is the dissipation stage of the thunderstorm.

Updrafts in clear air range from a few centimeters per second to more than 30 m per second. The largest updrafts in nature are found in thunderstorms (along with freezing temperatures and liquid water—remember airplane icing!). So it is easy to see why this is no type of weather in which to fly.

Remember that a thunderstorm is a consequence of atmospheric instability, an overturning of air layers in order for the atmosphere to achieve a more stable density stratification. Moisture is one of the atmospheric properties necessary for the formation of thunderstorms, and moisture is usually brought into the area by convection of moist air from a body of water. An example is the Gulf of Mexico; when moist air tongues flow over the Midwest and Mississippi Valley, thunderstorm possibilities are created.

The lifting action that initiates thunderstorms can be either thermal or mechanical. Thermal lifting occurs when the surface is heated by insolation, which in turn heats the layer of air and initiates buoyancy. This causes the air to rise, developing a local circulation pattern known as convective lifting. Storms begun in this manner are called air-mass thunderstorms. These convective systems of cumulonimbus clouds range from isolated turrets of only a mile or so in diameter to great masses of thunderstorms several miles in diameter which are seen in dramatic long-distance snapshots.

The first and most famous scientific experiment to show lightning as a form of electricity was conducted by Benjamin Franklin in 1750. Using a kite and key, he demonstrated that lightning was an electrical discharge. After his remarkable discovery, he proceeded to invent the lightning rod. Since that time, thunderstorms have been studied as a scientifically explainable phenomenon rather than a supernatural manifestation. The fact that many investigators have lost their lives in exploring the electrical processes associated with its violent air currents accounts in part for the slow progress in the understanding of thunderstorms and also attests to the violence of this phenomenon.

One of the earliest descriptions of the turbulent motions and hazardous precipitation of a thunderstorm was provided 93 years after Benjamin Franklin's discovery, by Professor John Wise of Carlisle, Pennsylvania, who was swept into one in a balloon. In his vivid description of his junket into a cumulonimbus cloud he provides an excellent portrayal of what goes on in what appears to be a limited-state thunderstorm:

According to announcement, I started on Saturday last on my forty-first serial excursion from the Center Square of Carlisle, Pa. at precisely fifteen minutes past two o'clock in the afternoon, it being the 17th of June, 1843. A slight breeze from the west wafted me a short distance in its direction horizontally, after which the ascent became nearly perpendicular until the height attained was about twenty-five hundred feet, when the balloon moved off toward the east with a velocity much greater than that of its ascent. When I had reached a point about two miles east of the town, there appeared a little distance beyond and above me, a huge black cloud. Seeing that the horizontal velocity of the balloon would carry it underneath and beyond the cloud, preparations were at once made to effect it by throwing out some ballast as soon as its border should be reached. Harrisburg was now distinctly in view, and the balloon moving directly for it; I was hesitating, with the bag of ballast in my hand, whether I should throw it out for the purpose designated, or continue straight on as I was then going to the place just mentioned. By this time I had reached a point underneath the cloud, which was expanding, and immediately felt an agitation in the machinery, and presently an upward tendency of the balloon, which also commenced to rotate rapidly on its vertical axis. I might have discharged gas and probably have passed underneath it, but thinking that it would soon be penetrated, and then might be passed above, as it appeared not to be moving along itself, I made no hesitation in letting the balloon go on its own way. . . . The cloud, to the best of my judgment, covered an area of from four to six miles in diameter; it appeared of a circular form as I entered it, considerably depressed in its lower surface, presenting a great concavity toward the earth, with its lower edges very ragged and falling downward with an agitated motion, and it was of a dark smoke color. Just before entering this cloud, I noticed, at some distance off, a storm cloud from which there was apparently a heavy rain descending. The first sensations I experienced when entering this cloud were extremely unpleasant. . . . The cold had now become intense, and everything around me of a fibrous nature become thickly covered with hoar-frost, my whiskers jutting out with it far beyond my face, and the cords running up from my car [the balloon basket] looking like glass rods, these being glazed with ice, and snow and hail was indiscriminately pelting all around me. The cloud, at this point, which I presumed to be about the midst of it from the terrible ebullition going on, had not that black appearance I observed on entering it, but was of a light, milky color, and so dense just at this time that I could hardly see the balloon, which was sixteen feet above the car. From the intensity of the cold in this cloud I supposed that the gas would rapidly condense, and the balloon consequently descend and take me out of it. In this, however, I was doomed to disappointment, for I soon found myself whirling upward with a fearful rapidity, the balloon gyrating and the car describing a large circle in the cloud. A noise resembling the rushing of a thousand milldams, intermingled with a dismal moaning sound of wind, surrounded me in this terrible flight. Whether this noise was occasioned by the hail and snow which were so fearfully pelting the balloon, I am unable to tell, as the moaning sound must evidently have had another source. I was in hope, when being hurled rapidly upward, that I should escape from the top of the cloud; but as in former expectations of an opposite release from this terrible place, disappointment was again my lot, and the congenial sunshine, invariably above, which had already been anticipated by its faint glimmer through the top of the cloud, soon vanished, with a violent downward surge of the balloon, as it appeared to me, of some hundred feet. The balloon subsided, only to be hurled upward again, when, having attained its maximum, it would again sink down with a swinging and fearful velocity, to be carried up again and let fall. This happened eight or ten times, all the time the storm raging with unabated fury, while the discharge of ballast would not let me out at the top of the cloud, nor discharge of gas out of the bottom of it, though I had expended at least thirty pounds of the former in the first attempt, and no less than a thousand cubic feet of the latter, for the balloon had also become perforated with holes by the icicles that were formed where the melted snow ran on the cords at the point where they diverged from the balloon, and would by the surging and swinging motion pierce it through.

. . . Once I saw the earth through a chasm in the cloud but was hurled up once more after that, when, to my great joy, I fell clear out of it, after having been belched up and

swallowed down repeatedly by this huge and terrific monster of the air for a space of twenty minutes, which seemed like an age, for I thought my watch had been stopped, till a comparison of it with another afterward proved the contrary. I landed, in the midst of a pouring rain, on the farm of Mr. Goodyear, five miles from Carlisle, in a fallow field, where the dashing rain bespattered me with mud from head to foot as I stood in my car looking up at the fearful element which had just disgorged me.

The density of this cloud did not appear alike all through it, as I could at times see the balloon very distinctly above me, also, occasionally, pieces of paper and whole newspapers, of which a considerable quantity were blown out of my car. I also noticed a violent convolutionary motion or action of the vapor of the cloud going on, and a promiscuous scattering of the hail and snow, as though it were projected from every point of the compass.*

A more recent investigation of thunderstorms was made in the late 1940s, by the Meteorological Group at the University of Chicago under the direction of Professor Horace Byers. This was the most extensive investigation of the thunderstorm and related phenomena ever undertaken. The Chicago group discovered that the thunderstorm undergoes mutagenesis in three distinct stages during its life cycle. For cumulonimbus to develop, the atmosphere must possess: (1) sufficient water vapor, (2) an unstable atmosphere, and (3) a lifting action to initiate the essential vertical development. When these attributes are present, the initial cumulus blooms into a cell with many similar clouds to form a cluster of cumulus clouds. If this is to become a cumulonimbus cloud with great vertical development and evolve into a thunderstorm, it must have updrafts of greater magnitude and be more uniform than those cumulus that do not become thunderstorms. The cumulus stage or "youthful period" of the cumulonimbus is described as its growing period, in which the clouds are cumulus congestus, cumulus pileus, or cumulonimbus calvus type. The convection at this stage of development consists primarily of updrafts that increase in magnitude as they penetrate the higher altitudes (see Figure 6-1). These strong vertical motions often attempt to force the growing cloud tops to pierce the tropopause. Most cumulonimbus clouds never

---

*From Taylor, George F. *Elementary Meteorology*. Englewood Cliffs, N.J.: Prentice-Hall, Inc., © 1954, pp. 225-226. By permission of Prentice-Hall, Inc.

intrude upon these extreme heights, but they attain altitudes of somewhere between 7 and 15 km (25,000–50,000 feet) in the middle latitudes. When the full-fledged cumulonimbus clouds become a thunderstorm, they may reach very high levels, since they are nearing the termination of their youth or cumulus stage. During their period of growth, these clouds of vertical development are associated with strong updrafts. These, in their turn, entrain the unsaturated environmental air, as it mixes with the converging air of the surface to feed the updrafts, with everything transpiring as the higher levels are penetrated by the growing cell.

Pilots who have encountered the cloud at this stage have noted that the temperatures within the cloud in the developing stage are warmer than those of the environment at corresponding altitudes. Even though hydrometeors of both rain and snow are present during the latter part of this stage, they remain suspended in the cloud. Strong updrafts cause the rain droplets and/or snowflakes to remain suspended in the cloud rather than fall as precipitation.

The second or mature stage of the thunderstorm is characterized by updrafts and the beginning of downdrafts, especially in the lower portion of the cell. Precipitation (usually rain) begins to fall from the base of the cloud. The rain or snow from these storms usually reaches the ground except in arid regions, where it becomes virga (precipitation that evaporates prior to reaching the ground). If virga occurs, we have a dry thunderstorm. Such storms can cause many wildfires and constitute a hazard in the western United States and other arid areas of the world.

The maturing process of the thunderstorm begins when the water droplets have grown by the process of coalescence, until they reach a critical mass that can no longer be sustained in the cloud by the updrafts. It is now that the hydrometeors fall from the base of the cloud and become precipitation. The water drops falling as precipitation with their large masses exert frictional drag within the updraft, changing the updraft into a downdraft. This process occurs at about the time the first precipitation reaches the ground. The downdraft is reinforced by two processes:

1. The downdraft will be accelerated downward because the air of the thunderstorm is unstable.
2. The air moving past the falling hydrometeors evaporates some of the liquid moisture, which cools both the droplet and the adjacent air.

Thus, precipitation initiates the demise of the thunderstorm. It should be noted that not all of the updrafts become downdrafts; those updrafts that are not in the precipitation cells continue and attain their greatest magnitude early in the mature stage near the top of the clouds. The downdraft reaches its maximum amplitude in the lower parts of the cloud system. Downdrafts have attained speeds of 60 knots in the forward portion of a line of thunderstorms. These high-speed gusty winds have been called cold air avalanches by some observers, and they are found near areas of precipitation and divergence. The temperatures found in the updrafts are often warmer than the temperatures of the cloud environment at similar levels. Also, the temperatures of the downdrafts have been found to be colder than those of the environment.

The mature stage supports the most intense thunderstorms, with their associated pyrotechnical display, severe to extreme turbulence, hail, and severe icing conditions. On the ground, we find heavy rains, strong and gusty winds, and hail—just the kind of weather that makes you feel like saying, "I'd rather stay in bed." The clouds of the mature-stage thunderstorm are the ones that may penetrate the tropopause and enter the lower stratosphere. When a storm reaches this stage, the downdrafts expand both vertically and horizontally to encompass the entire cell. This signals the commencement of the dissipating stage. A terminal stage occurs when the moisture source for the updrafts, with its associated latent heat supply, is cut off. The thunderstorm at that point has entered its final hour. Professor Byers's group discovered that as the moisture source is removed, the thunderstorm dissipates its energy, precipitation ceases, and the vertical motion is terminated because the temperature difference between the convective cell and its environment is nil. The cumulonimbus clouds that compose the cells of the thunderstorm ultimately vanish, leaving only a few stratus-type clouds indicating that the atmosphere is now stable and safe for flying. That is, until the next thunderstorm appears on the horizon, which poses some questions: (a) Is it a single storm or is it one in a line of thunderstorms? (If it is a loner, an isolated turret only a mile or so in diameter, it is called a limited-state or air-mass thunderstorm.) (b) Is it instead only one of a long line of thunderstorms? (Then it is referred to as a steady-state thunderstorm, which is associated with an organized weather system, that is, a frontal or squall line.)

The limited-state thunderstorm is less feared by aviators

because it usually is a loner and does not have the organized upper air support found associated with the steady-state variety. The limited-state thunderstorm is usually rather easy to avoid. It is not usually a major undertaking to fly around its perimeter, unless you're hang gliding or soaring. Remember that *any* thunderstorm can belch hail as far as 5 miles from it. These loners are not the most violent of storms, but they can be treacherous to anyone who imagines they are harmless and attempts penetration in an aircraft. Just recall the encounter that Professor John Wise had with a single lonesome cumulonimbus cloud! Surely such an experience would convince even the most stubborn pilot that even an isolated thunderstorm should be avoided. Air-mass thunderstorms are primarily an afternoon phenomenon. This is because the lifting action that initiates their activity is thermal: The heated surface warms the adjacent moist atmosphere, which expands and rises into conditionally unstable air, which develops into an air-mass thunderstorm commencing its life cycle as a little monster loosed in the atmosphere. *Any* thunderstorm can produce a degree of turbulence that can cause structural damage to aircraft.

The steady-state thunderstorms have two basic subdivisions: (1) the frontal thunderstorm, and (2) the squall-line (instability) thunderstorm. Frontal thunderstorms occur in moist, warm, unstable air that is mechanically lifted by a frontal surface. Cold air pushes under warm, moist, unstable air and triggers vertically rising air currents that develop into cumulonimbus clouds and result in thunderstorms. Less commonly, warm, moist, unstable air glides over cold, stable air. The storms form in a line along the frontal surface. Squall-line thunderstorms are not directly associated with frontal weather and have a pronounced inclination to form in bands or lines along the direction of winds at low levels in the warm sector of a cyclonic system. These are found between 50 and 300 miles ahead of the cold front, usually parallel in orientation. The lifting mechanism is not completely understood, but several theories have been suggested. The trough aloft might create an area or region of convergence at the surface to initiate vertically moving air in the warm, moist, conditionally unstable air mass. This would produce a line of thunderstorms in the warm sector of the cyclone. Or a hydraulic jump or pressure jump might be the mechanism that initiates the lifting and launches the accelerating updrafts that develop into cumulonimbus and thunderstorms.

The downdrafts associated with mature thunderstorms often travel miles ahead of the thunderstorm, and this outflow of cold air is commonly called a pseudo-cold front, residing in the warm, moist sector of the cyclone. This mass of cold air moving in the warm air can cause lifting and initiate the formation of a squall line ahead of the frontal system.

It is not often that a limited-state thunderstorm will last for a very long time, except in the arid mountainous regions of the western United States. There is a strong tendency for cumulus cells to form and join together with adjacent cells by an interconnected cloud structure. These interconnected clouds are recognized by pilots as zones of less turbulent air, and are usually seen as a separate pattern of cloud radar echoes.

Now that the thunderstorm has been defined, you can better appreciate why you must do so much planning to make a decision involving flying in and around them. As you examine where and when thunderstorms occur most frequently, explaining to yourself the causes and natural history of this weather phenomenon, you will learn to avoid them in preflight planning and how to take early, calm, evasive action in flight without actually encountering them.

The frequency of thunderstorms varies throughout the world. In the tropical regions, thunderstorms occur throughout the year. In the middle latitudes, they occur most frequently in the springtime, occasionally in the summer and fall, and less often in the winter. The arctic regions occasionally experience thunderstorms during the summer months. The annual graphical distribution of thunderstorms in the contiguous 48 United States shows that few occur along the Pacific coast of the U.S. (less than 5). The greatest occurrence is in Florida (more than 90) and along the Gulf coast (70 per year). There were 60 in northeastern New Mexico and 40 in West Yellowstone, Montana. Remember that in order to have thunderstorms, three ingredients are required: (1) atmospheric instability, (2) a supply of sufficient moisture, (3) a lifting action. Florida and the Gulf coast have an abundant supply of atmospheric moisture that migrates northward toward the Great Plains, Mississippi Valley, and Appalachian Mountains in the eastern United States, and to the High Plains and Rocky Mountains in the western United States. This warm, moist air is conditionally unstable. Thunderstorms are then initiated whenever a lifting occurs and such air is present. The lifting can be frontal, such as when an arctic or polar cold front moves southward from Canada or a Pacific cold front advances southeastward from the Rockies. The lifting can also be initiated by solar heating of the surface in the afternoon, which is quite frequent during the spring and summer months.

Also, as the maritime, moisture-laden air from the Gulf climbs the Appalachian Mountains to the east or the Rocky Mountains to the west, thunderstorms are formed as orographic lifting instigates their development. During the latter part of July and August, moist air from the Gulf even penetrates into the Great Basin, bringing both convective and orographically spawned thunderstorms. The Gulf moisture often penetrates as far north as the prairie provinces of southern Canada. The thunderstorm occurrence varies widely from season to season, but generally they are most frequent during the spring and summer seasons.

Thunderstorms don't just happen! The unstable moisture-laden atmosphere needs a lifting action to initiate these awesome storms. The mechanical lifting by orographic or frontal systems has been mentioned. We've also discussed the mechanism of convective lifting; how does it function? One way is surface heating, which creates buoyancy on warm afternoons when the sunshine heats the ground surface, which then warms the air immediately over it. As the adjacent air is warmed, it expands, becoming less dense and rising as the buoyant forces acting on it become more active. If the lifting occurs when the air is conditionally unstable and moisture laden, cumulus clouds will form and soon grow into cumulonimbus cells and thunderstorms. This is especially common over uneven and mountainous terrain, but may be found over any type of surface. Another cause of vertical motion is initiated when surface winds converge. This effect is caused by local topography and uneven solar heating of the surface and its influence on the patterns of local circulation. Again, lifting results from these flow patterns of the air when moisture is present and the air is unstable.

Radar can assist aircraft in avoiding thunderstorms. Don't, however, place all confidence in radar when considering how to fly a safe course around a thunderstorm or a line of thunderstorms. It has been shown that raindrops reflect radar signals, but not necessarily the cloud droplets. Meteorologists have shown that drop size is directly proportional to the rate of rainfall, and the heaviest intensity of precipitation is associated with thunderstorms. This suggests that the strongest echos are associated with thunderstorms. One excellent use of airborne radar has been to assist aviators in avoiding thunderstorms.

Professor Wise's encounter with a limited-state thunderstorm revealed that just about every weather hazard known to aviation is present in one vicious package. These hazards accompanying a thunderstorm occur either singly or in morbid combinations: (1) turbulence of all intensities, (2) icing and freezing precipitation in its various varieties, (3) low ceilings and poor visibilities, (4) blinding lightning flashes, (5) extreme altimeter fluctuations, (6) hail, (7) precipitation static, and (8) thunderstorm electricity. The smart pilot knows that these hazards are easier to avoid than fight.

Tornadoes, the most violent of meteorological phenomena, develop from severe thunderstorms. These storms draw the air into their bases with great vigor. The rapidly upward-moving air will create a low-pressure cell near the surface. Often, the low-pressure area will generate an initial rotational motion that may form an extremely concentrated vortex extending from the earth's surface into the mother cloud. The winds and associated dust that form the visible vortex have been estimated to exceed 200 knots at times and are balanced by exceptionally low pressure at the center of the vortex. Estimates have been made by some investigators that the central pressure of the vortex has been reduced by as much as one-third of the atmosphere. These central pressures are so low that most of the structural damage to buildings is explosive, resulting from the sudden lowering of the atmospheric pressure. As the storm vortex passes over buildings, a tremendous pressure force is exerted on windows, outside walls, and roofs of buildings. This short-lived pressure difference reduces some buildings to debris.

Studies have shown that of nearly 200 tornadoes per year occurring in the United States, about 90 percent are associated with squall lines (which are 150 miles ahead of the cold front in the warm-air sector). Most tornadoes occur in the central United States and along the Gulf coast. An aircraft entering a tornado vortex will at least sustain structural damage, if not be broken apart. Since the vortex cloud is known to extend well into the mother cloud, any pilot caught on instruments in a severe thunderstorm could, at any time, encounter a hidden vortex. Goodbye to that pilot! It has been found that tornadoes often occur in families, because if conditions are present for the formation of one tornado, then they are present so that many tornadoes can form. Cumulomammatus clouds are present when violent thunderstorms occur. Any cloud that displays rounded, irregular pockets and bulges extending from its base is a red flag that tells pilots severe turbulence is also present. Aviation weather reports specifically mention these hazardous clouds.

Turbulence associated with thunderstorms results from vertical and horizontal air motions, gusting winds, wind shears, and so on. Turbulence is always associated with thunderstorms. Turbulent air travels in various directions at various speeds and occurs in organized updrafts and downdrafts, each having strong vertical and horizontal wind shears. Turbulence found in thunderstorms varies in intensity from moderate to extreme. Even outside the cumulus clouds, wind-shear turbulence has been encountered from several thousand feet above the clouds up to 20 miles laterally away from the storm! A severe local storm is always especially dangerous. Low-level turbulence in the shear zone between downdrafts and the surrounding air often develops into a roll cloud along the leading edge of the thunderhead. The roll cloud is similar to the cloud associated with a mountain wave since it also generates a great amount of severe to extreme turbulence.

If the pilot inadvertently penetrates a thunderstorm, he will find that the stresses on the aircraft will be least if the aircraft is held at a constant attitude and allowed to ride the drafts. Always remember there is no sure way to pick the soft spots of thunderstorms and safely fly through them. Another hazard to flight in and about thunderheads is *icing*. The updrafts in the cumulonimbus clouds provide an abundant amount of liquid water that is carried to heights above the freezing level, where the water becomes undercooled. Anytime undercooled water is present at any level in the atmosphere and a moving foreign object comes in contact with it, the water freezes immediately and adheres to that object. Icing can occur at any altitude above the freezing level. An abundance of undercooled water makes icing especially hazardous, and this is most frequent between the temperatures of 0°C and -15°C. Thus, all of the ingredients are present for the occurrence of icing in thunderstorms. Icing is even more hazardous when it is accompanied by turbulence. Still another hazard found in and around thunderstorms is hail. Hail and hailstones can pound an aircraft mercilessly, and alter the airfoil (which in turn causes the aircraft to lose lift), crack windows, etc. Hail has been observed in clear air several miles from its thunderstorm source.

Other hazards to flight associated with thunderstorms are low ceilings and poor or reduced visibilities. The visibility may be nil (zero) in the cumulonimbus cloud and greatly reduced in the precipitation beneath the cloud. At times dust is lifted by the first gust generated by the thunderstorm. These hazards are magnified when the other hazards of thunderstorm flying are considered. Still another is lightning, which becomes a problem when the bright flashes temporarily blind the pilot. This happens because the flashes are so bright that they provide a marked contrast to the dark interior of the clouds. If you are ever unlucky enough (or foolish enough) to be caught in a thunderstorm, always turn the lights in the cockpit and those of the instrument panel to their *brightest* setting. The lightning flashes will then have the least effect on your vision, because your eyes will be used to the contrast.

In addition to brightness, lightning presents an additional hazard—electricity! Lightning can strike and puncture the skin of an aircraft, which will cause it to lose pressurization and damage the electronics of the communication and navigational equipment. It also has been suspected of igniting fuel vapors and causing onboard explosions. However, serious accidents due to lightning strikes are rare.

Let us examine still another problem that may be encountered in flying merely too close to thunderstorm clouds. This is the negative effect of sudden pressure variations on altimeters and also rate-of-climb indicators. Altimeters may have an error of over 100 ft in a 15-minute period during a flight near a thunderstorm. The rate-of-climb indicator may also give false readings, which can cause an inexperienced pilot to panic and lose self-confidence in his ability to cope with these situations.

A good rule of thumb to follow when flying around and near thunderstorms is this: "The more frequent the lightning, the more severe the thunderstorm." Increased frequency of lightning indicates that a thunderstorm is increasing in intensity, and a decreased frequency suggests a decreasing intensity of the thunderstorm. At night, if you see lightning playing along the horizon, remember that this could be an indication that you are approaching a squall line or a line of frontal thunderstorms. Still another hazard is precipitation static, a steady high-level noise in radio receivers which is caused by the intense corona discharges from sharp metallic objects and the edges of an aircraft in flight in the vicinity of these electrical storms. Aircraft flying through clouds, precipitation, or a concentration of solid particles of ice or dust will accumulate a charge of static electricity which may discharge into the air, causing noisy disturbances at the lower frequencies. These luminous discharges are visible at night; although they seem rather eerie, they are really harmless. This phenomenon was named *Saint Elmo's fire* by Mediterranean sailors who saw these bushy discharges on the top of ship masts. On propeller aircraft, Saint Elmo's fire has been known to illuminate the tip of the propeller as an eerie, firelike brilliance; it has also been known to cause some pilots to turn a bit pale.

The smart pilot will always avoid all thunderstorms, because it is far easier to get into trouble than it is to get out of it! One of the better ways to avoid trouble is to make your decision about which route and alternatives to take before takeoff. And if the storm conditions are all around you while you are still on the ground, you simply do not take off at all! Stay home and read a good book on aviation meteorology!

7 TURBULENCE

Turbulence can be anything from annoying bumpiness to violent tossing, which can subject an aircraft to structural damage and make controlled flight impossible. It occurs in many different types of weather situations, even in clear skies. Turbulence may be isolated in location or cover an extensive area. It develops from irregular shifts or eddies in the air, creating gustiness in any and all directions. Especially a hazard during takeoffs and landings, when the air speed is close to the aircraft's critical stalling speed, turbulence is also found at all levels of flight, even in the high atmosphere. Eddies causing bumpiness are about the same order of magnitude as the size of an aircraft. Eddies occur in an irregular pattern that makes forecasting of the specific eddy location next to impossible. The reaction of turbulence varies not only with the frequency and intensity of the irregular motions of the atmosphere, but also with the characteristics of the aircraft, including the air speed, wing loading, and attitude. Since some knowledge of turbulence is necessary for the pilot because of the hazard it presents to flight, we will now examine methods by which turbulence is classified. It is difficult for the forecaster to predict the occurrence, location, and intensity of turbulence. The pilot's judgment of the intensity of turbulence may be influenced by the length of time his plane is subject to it. Obviously a brief encounter will not be considered as significant as one lasting for several minutes. A pilot's previous experience with turbulent conditions and its effect on a given aircraft type will also affect any evaluation of intensity. Forecasters have *no direct measure* of turbulence. In fact, the only time that a forecaster knows turbulence occurs is when a pilot encounters and reports it. The forecaster does not know when and where turbulence will occur, but he has a good general knowledge of the terrain features and meteorological indicators that are usually associated with turbulence; but these meteorological indicators are only available at infrequent intervals and at widely separated locations. The pilot report is one of the most valuable tools the forecaster has in predicting turbulence aside from cumuliform clouds, so let us examine turbulence reporting criteria for the four classes of turbulence: light, moderate, severe, and extreme.

Light turbulence momentarily causes a slight erratic change in attitude or may cause a slight or rapid or similar rhythmic bumpiness without an appreciable change in attitude or altitude. The occupants of the aircraft may feel a slight strain against their seat belts or shoulder straps. Unsecure objects may be slightly displaced. Food service may be continued with no difficulty.

Moderate turbulence is similar to light turbulence but is of greater intensity. Changes in altitude and/or attitude usually cause large variations in indicated air speed. Moderate turbulence is similar to that of light shock but of greater intensity. It causes air bumps or jolts without appreciable change in the aircraft's altitude or attitude. Occupants feel definite strain against the seat belts and shoulder straps. Unsecured objects are dislodged. Food service and walking are difficult.

Severe turbulence causes abrupt changes in altitude and/or attitude and usually causes large variations in indicated air speed. The aircraft may momentarily be out of control. Occupants are forced violently against their seat belts and shoulder straps. Unsecured objects are tossed about. Food service and walking are impossible.

Extreme turbulence causes the aircraft to be violently tossed about and is practically impossible to control. It may cause structural damage.

Turbulence can also be divided into four general types based on meteorological and other properties responsible for producing it. These are convective, mechanical, wind shear, and high-level clear-air or wave-induced turbulence. One cause of clear-air turbulence (CAT) is wind shear in regions of large amounts of eddy kinetic energy near the jet stream and also in the vicinity of mountain waves. Clear-air turbulence is a special category, mostly because of its significance for jet and turbojet aircraft, and will be discussed later.

The most common type of turbulence is convective, which results from nonuniform heating of the earth's surface. *Convective or thermal turbulence* is common on sunny afternoons. The heated air rises in convective currents of variable strength. The upward moving air continues to accelerate until it reaches a level where the temperature of the rising air is equal to that of its environment. Pilots know that a smooth flight may be obtained if the aircraft can operate at an altitude above the ground-generated convective currents. It should be noted that both updrafts and downdrafts exist in regions of convective turbulence. The turbulence becomes more severe as the spacing becomes closer between the updraft and downdraft. Even when the air is dry and no cumulus clouds are present, convective turbulence may be found. Thunderstorms create the ultimate in convective turbulence, as discussed in Chapter 6.

With regard to *mechanical turbulence*, it should be noted that if an object is in a moving fluid (such as air), it impedes the flow. This causes a directional change in the path of the flow. After the current passes the object, it seeks to return to its original flow pattern, and eddy currents are created on the lee side of the obstruction. These eddies are manifestations of mechanical turbulence. When turbulence results from an obstruction in the path of wind and not from any of the various meteorological parameters, this turbulence is called mechanical. The degree of intensity of mechanical turbulence depends on meteorological parameters. Some natural objects are conducive to the formation of mechanical turbulence: mountains, hills, buildings, and even moving aircraft. Heavy aircraft create considerable turbulence, as is evidenced by the wake turbulence generated by large transport aircraft. In those locations where obstacles block the path of air flow, the normal wind patterns become a complicated snarl of eddies. These can be likened to the rapids of a fast-flowing mountain stream but may be much more complex, since the eddies of the atmosphere have a much greater vertical magnitude. These eddies are carried along in the flow pattern of the wind. The size and extent of the eddies will have considerable effect on the flying characteristics of the aircraft. The intensity of the turbulence that an aircraft will encounter when flying (over rough terrain) will depend on the speed of the wind, the roughness of the surface, and the stability of the air. Low-level mechanical turbulence is extremely important in the takeoff and landing procedure, especially for light aircraft. Gusty winds have been known to cause many aircraft accidents and/or incidents. It is important for a pilot to be alert for the occurrence of turbulent eddies in the vicinity of hangars and other buildings located near the runway during takeoff and landing. If the wind speed is light, eddies will be created and tend to remain in rotating pockets of air near the buildings. However, if the wind speed is greater than 20 knots, the flow pattern will be broken into irregular eddies that may be carried downstream and can create hazards in the landing area.

Turbulence may be of minor significance in air moving slowly over a smooth surface of gently rolling hills. In such cases, its effect is most often felt within a few hundred feet of the ground. When the winds are blowing faster and the obstructions are larger, turbulence will increase in intensity and extend to higher levels in the atmosphere. High-level mechanical turbulence occurs when the wind blows over and perpendicular to a high, large mountain range. On the windward slope, it is usually rather smooth if the air is stable. On the leeward side, the wind spills rapidly down the slope, and strong downdrafts are generated. This causes the air to be very turbulent, as is shown in Figure 7-1. These downdrafts are very dangerous, and will tend to force

FIGURE 7-1 Mountain wave.

an aircraft into the ground if the aircraft is captured by the downdraft. Pilots should fly with caution when flying near mountain ridges during strong wind conditions, and avoid such hazardous areas when flying over this terrain. These areas are especially hazardous to flight when frontal or cyclonic activity dominates the weather in the vicinity of the mountains. These winds usually prove to be more turbulent and much stronger than those found in the general flow pattern. The winds in and near the mountains generally flow in the direction of passes and valleys. This will increase the air velocity through any narrow pass (canyon winds).

When winds are in excess of 50 knots, and are blowing approximately perpendicular to a high ridge, the resulting turbulence may be extreme. It is under these conditions that large waves tend to form on the leeside of the mountains. These waves have at times tended to extend upward to the tropopause or even higher. In horizontal dimensions they sometimes extend as far as 100 miles downstream from the mountain ridge. Figure 7-1 shows that when air blows over a mountain ridge and has ample water vapor present, characteristic orographic pendant clouds form over the ridge. Also, the altocumulus standing lenticular cloud, a rotor cloud, will form under the mountain crest. The standing-wave and rotor cloud are sure signs of severe to extreme turbulent conditions but severe turbulence can be present even when these characteristic clouds are not. The lenticular clouds are indicative of standing or mountain waves. When sufficient moisture is present, standing mountain waves with rotor clouds will be reported by pilots, generally associated with severe to extreme turbulence. Probably, the most dangerous aspect of a standing mountain wave is the magnitude of the sustained vertical motions. These clouds are rather common on the east side of the Sierra and Rocky Mountains. Mountain waves are not uncommon to the relatively low Appalachian Mountains but occur less frequently there than in the Rockies. Standing waves should be anticipated to be on the lee side of major mountain chains when the wind is blowing perpendicular to the ridge and exceeds 50 knots. Because of the lee-side pressure variation, some pilots estimate that altimeter readings are as much as a few hundred feet in error during turbulent encounters in mountainous areas.

Another form of mechanical turbulence is that generated by an aircraft. This turbulence is referred to as *wake turbulence*, and results as the normal air flow is disturbed and displaced by the aircraft as the plane quickly moves through the air. The wake turbulence generated by general-aviation aircraft is usually rather light. For some of the larger transport-type flying machines, however, wake turbulence can be considerable and outright hazardous to small craft taking off or landing in the wake of large aircraft. Wake turbulence develops because an aircraft obtains its lift by accelerating a mass of air over its airfoils. Thus, whenever the wings provide lift to sustain flight, air must flow over the wings. This generates a rotary motion or vortex of air extending from the wingtips. The landing gear, while bearing the entire weight of the aircraft prior to takeoff and after landing, may prevent the vortices from developing. But the instant a pilot in the takeoff roll pulls back the controls, these vortices are initiated. They may appear behind the plane as it lifts off. The vortices continue throughout the flight until the aircraft is again firmly settled on the ground. They spread *downward* and *outward* from the path of flight and drift with the wind. The strength of the vortices is proportional to the weight of the aircraft and the wing loading, so wake turbulence is more intense behind large transport aircraft than behind the small aircraft. Generally, it is a problem only when following large aircraft on the runway or crossing behind its path of flight. Turbulence usually persists for several minutes but may linger longer along the flight path. Airport tower controllers will inform pilots of possible wake turbulence, but you are in command of your own aircraft, and its safety is *your* responsibility. Most transport jets lift the nosewheel about midpoint down the runway in the takeoff run. Therefore, wingtip vortices begin as the nosewheel lifts off at about the middle of the takeoff run. The vortices behind propeller-driven aircraft begin only a short time prior to liftoff. During landing of either type of aircraft, the vortices end at about the point on the runway where the nosewheel touches down.

Pilots of small craft should avoid flying through these vortices when using the same runway as heavier aircraft. When landing behind a heavier aircraft, your approach must be above that of the landing aircraft and the touchdown made beyond the point where its nosewheel touched down. If landing after a departing aircraft, land only if the landing roll can be completed before reaching the midpoint of the takeoff run. If following a departing aircraft, take off only if you can become airborne before reaching the midpoint of the other plane's takeoff run, and then only if you can climb fast enough to stay above its flight path. If departing behind a landing aircraft, taxi onto the runway beyond the point where the nosewheel of the landing aircraft touched down, and make sure you have sufficient runway for a safe takeoff.

If parallel runways are available and a heavier aircraft takes off with a crosswind on the downwind runway, you may safely use the upwind runway. *Never* land or take off downwind when a heavier aircraft is using the upwind runway. When using a runway crossing another runway, you may safely use the upwind portion of your runway. You may cross behind the departing aircraft, if you stay behind the midpoint of its takeoff. Only cross after a landing aircraft if you are ahead of the point on the runway at which its nosewheel touched down.

If none of these situations are possible, wait at least 5 minutes or so for the vortices to dissipate or blow from the runway before you take off. These procedures are elementary, and the problems of wake turbulence are more operational than meteorological. The FAA issues periodic advisory circulars of such operational problems, so if you operate out of airports used regularly by air carriers, you should keep current on the latest advisories on wake turbulence and other operation problems.

As mentioned previously, wake turbulence is a form of mechanical turbulence created by flying aircraft, and can be very hazardous. The larger aircraft actually generates wingtip vortices (within the magnitude of small tornadoes!) which develop severe to extreme turbulence. For this reason, the FAA regulations require that transport-type aircraft be given clearance of *at least* 6 miles. These vortices have been known to move in every direction from the mother plane.

Wind-shear turbulence is located in the shear zone between two strata of differing wind velocity. Turbulence increases as the wind shear increases, and can be present almost anywhere in the atmosphere. A narrow zone of wind shear and its accompanying turbulence are often encountered when climbing or descending through a temperature inversion. Wind speed and direction sometimes change very abruptly with altitude in this zone. This danger is most intense when a strong temperature inversion exists. It presents a hazard to aircraft immediately after takeoff and on the final approach for landing. Aircraft entering this zone may experience extreme turbulence. This condition is not uncommon on the central and high planes and in the valleys of the Rocky Mountains. When severe turbulence is encountered just a few hundred feet above the ground, conditions are much more dangerous than if it is encountered where sufficient reaction room exists to overcome the problem. One of the reasons pilots are so very careful when landing or taking off is that meteorological problems are most dangerous at this stage of a flight. Remember, whenever you encounter turbulence while flying, always send a *pilot report*, so that any other planes that happen to be in your area can avoid that particular situation. Had you a pilot report telling you of the turbulence you just encountered, you could have avoided it by altering your particular flight path. Remember to help your fellow pilots and help the weather service by sending pilot reports.

There is one other class of turbulence, clear-air turbulence (CAT). Clear-air turbulence is associated with high-altitude flying at or near the tropopause adjacent to the jet stream. The name implies turbulence without the presence of clouds. Meteorologists of United Airlines have suggested that clear-air turbulence be called *wave-induced turbulence* (WIT) because it is associated with high-level waves of the atmosphere. These waves usually occur in conjunction with the jet stream as it skirts the polar air mass high above the path of surface cyclones and fronts. CAT has a preferred location near the upper trough on the poleward side of the jet stream. Another frequent location of CAT is along the jet stream north and northeast of a rapidly deepening surface low. It is also found on the lee side of mountain ridges when mountain waves are present. CAT associated with mountain waves has been encountered from mountain crests to as high as 5,000 ft above the tropopause, and as far as 100 miles downwind from the ridge! This suggests to the pilot that he should be aware of the occurrence of clear-air turbulence anywhere in the vicinity downwind of mountain waves. Remember, these are only *preferred* locations for CAT; it can occur anywhere and at times when there seems to be no reason for its existence. Some have theorized that strong winds carry the turbulent volume of air away from its source region. Generally when this happens, the intensity of turbulence diminishes downstream, but still some turbulence may be encountered where it normally would not be expected. Pilots will find that CAT forecast areas are sometimes elongated to indicate a probable downwind drifting of the turbulence from its source region.

A forecast of turbulence specifies a volume of air space that is smaller than the total volume of air space used by aviation. It is still a relatively large volume when compared to the localized extent of the hazard. Because the nature of the forecast volume is patchy, turbulence can be expected to be encountered only intermittently, or possibly not at all. Experience has shown that a flight through a volume of forecast turbulence, on the average, encounters only light and annoying turbulence 10 to 15 percent of the time, and about 2 or 3 percent of the time there is a need to secure all objects. The odds of encountering hazardous turbulence are about 1 in 500.

Much turbulence can be avoided by proper preflight planning. This means that you use, with the help of the pilot briefer, the upper-air charts and forecasts to locate the jet stream, wind shears, and areas of most probable turbulence. On those rare occasions when it is impractical to completely avoid an area of forecast turbulence, proceed as if your life depends on the caution you exhibit while flying through the danger area. It may! Those areas that must be avoided at all costs are areas where the vertical wind shear exceeds 6 knots per 1,000 ft or the horizontal shear is in excess of 40 knots per 150 miles. If you were to ever get caught in CAT associated with the jet stream, you should always climb or descend a few thousand feet, moving farther from the jet core. If you experience CAT not associated with the jet stream, it is best to change altitude, since you do not know the horizontal extent of the phenomena.

Again, whenever you as a pilot encounter turbulence, always file a pilot report so the next flight along that route can avoid the danger. It is also useful to make in-flight reports when no CAT is experienced, especially where it normally might be expected.

8 ICING AND IFR WEATHER

ICING   Aircraft icing is a major hazard to aviation. It can alter the flight characteristics of an aircraft until it is unable to fly, or it can choke the engine's fuel-induction system until sufficient power to maintain flight is not available. Information on icing is an important part of flight training. Icing is a *cumulative hazard* to aviation. First, we must ask ourselves, "What does icing do to an aircraft?" To answer our question, let us examine the cumulative effect of *structural icing* as is shown in Figure 8-1. Icing reduces the efficiency of the aircraft by (1) increasing the weight, (2) reducing the lift, (3) decreasing the thrust, and (4) increasing the drag. Each effect tends to increase the stalling speed of the aircraft. Icing also seriously lessens aircraft engine performance. It will adversely influence the operation of the flight instruments, disrupt radio communications, and impair the operation of the control surfaces, brakes, and landing gear.

    Because of these profound effects, pilots need a basic knowledge of icing, the seasonal and mountain effects of icing, and they must use this information wisely. Two conditions that are necessary for the formation and accumulation of ice on an aircraft are: (1) The aircraft must be flying through visible water such as drizzle, rain, or cloud droplets. (2) The temperature of the water must be below freezing (0°C or 32°F). The liquid water droplets being undercooled are unstable and quickly turn to ice

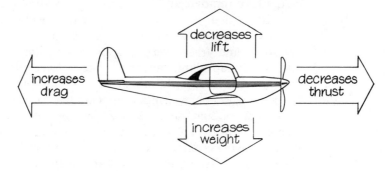

FIGURE 8-1 Cumulative effects of structural icing.

Clear - hard and glossy

Rime - brittle and frost-like

Mixed - hard rough conglomerate

FIGURE 8-2 Clear, rime, and mixed icing on airfoils.

when disturbed by an aircraft or other moving object. There are three categories of icing that pilots report: (1) clear, (2) rime, and (3) mixed. Each category has features that make it identifiable (see Figure 8-2).

Clear ice appears as a smooth glaze. It is formed by the slow freezing of large water drops such as those found in cumuloform clouds or in freezing precipitation. Accumulation rapidly increases as large drops spread over and assume the shape of the surface on which they freeze. The rather slow freezing rate of clear ice enables the air between the drops to escape, leaving the water to freeze as a clear, smooth, glassy ice formation. Clear ice is very difficult to remove, and for this reason is an immediate problem.

Rime ice is composed of small water droplets freezing quickly on exposed surfaces of the aircraft. Because the droplets are small and freeze very quickly, they retain their round shapes and the encapsulated air that rests between the droplets cannot escape prior to completion of the freezing process. This gives the ice a rough, coarse, milky appearance. It has the same texture as the frost that accumulates on the freezing coil of a home freezer without an automatic defroster. Frequently, it has an uneven surface. Rime is easily broken up and removed by deicing boots. It is associated most frequently with stratoform clouds and builds up less rapidly than does clear ice.

Mixed icing is present when drops vary in size from small cloud droplets to larger rain and shower particles, or when liquid drops are intermingled with snow and ice. Because of the rapid freezing, it forms quickly. It also can form when ice particles become imbedded in clear ice, building a rough accumulation on the leading edge of the airfoil. Mixed ice is a conglomerate of both rime and clear ice, and is very difficult for the pilot to remove with deicing equipment.

Now we shall examine the various icing intensities. There are four intensity classes: trace, light, moderate, and severe. Each is defined by the rate of ice accumulation. *Trace* icing occurs when ice becomes visible, which means that the rate of accumulation is slightly greater than the rate of sublimation. It is usually not a hazard, and deicing equipment is not utilized unless it is encountered over extensive periods of time, perhaps an hour or more. *Light* icing occurs when the rate of accumulation may create a problem if the flight is prolonged in a hostile environment for over an hour. A pilot occasionally can use deicing and/or anti-icing equipment to remove and prevent an accumulation. It does not present a problem if deicing or anti-icing equipment can be used. *Moderate* icing occurs when the rate of accumulation is such that even a short encounter becomes a potential hazard and the use of deicing or anti-icing equipment or diversion is necessary. *Severe* icing occurs when the deicing or anti-icing equipment fails to reduce the hazard. Immediate diversion is necessary for continued safe flight.

Let us examine the problems of aircraft icing. As we know, ice can form and adhere to the various exposed parts of the airplane. The principal effects are: (1) structural icing, (2) instrument icing, (3) carburetor icing, and (4) ground icing. All are really forms of structural icing, except carburetor icing, but they are considered independently in order to emphasize the dangers that each can cause.

Structural icing occurs most rapidly when flying through clouds composed of undercooled liquid water drops or freezing precipitation. As liquid water particles impact on the aircraft's exposed surface, they freeze and release part of their latent heat to the remaining water, permitting it to flow along the surface as a liquid before freezing. Ice that forms in this manner is clear ice. When clouds or precipitation contains a mixture of liquid water and ice particles, the frozen particles become imbedded in the clear ice, causing a rough, irregular accumulation, which can grow quite rapidly and is the most dangerous type of ice formation encountered by flying aircraft, mostly because of its irregular shape and excessive weight.

The rate of ice accretion on aircraft is affected by many factors: (1) the shape and size of the airfoil, (2) the speed of the aircraft, (3) the angle of the airfoil as it passes through the air. Some aircraft will tolerate more ice than others. All are impaired to some degree by the formation of ice on their superstructures. Although ice may form on any exposed surface of the aircraft, its formation on the airfoil and control surfaces presents the greatest danger to flight.

Airfoil icing includes not only the wing and tail surfaces, but also the propeller, since it is also a part of the airfoil system. Ice formation on the propeller greatly affects the power and speed of the aircraft. Another problem caused by propeller icing, besides reducing the aerodynamic thrust, is that fragments of ice may become dislodged and ejected from the propeller. This can cause it to become unbalanced, which may result in excessive vibration that can cause internal structural damage to the aircraft engine. Still another problem is that the ejected ice may become a projectile striking and inflicting damage upon some other part of the plane. Icing on the other surfaces is usually less hazardous to the performance of the aircraft, but nevertheless constitutes a serious danger to flight safety.

Instrument icing occurs when ice reduces the size of the inlet of the pitot tube. The reliability of the air-speed indicators is greatly reduced. Icing of the static inlet also reduces the effectiveness of the rate-of-climb and turn-and-bank indicators, which renders flying very precarious. Icing of the radio antenna may sever it from the aircraft and disrupt the communication and navigation systems.

Carburetor icing is the most frequent form of icing and is easily recognized and guarded against. Nearly all aircraft have carburetor heaters to prevent this problem. It occurs during the gasification of the fuel by the expansion process, which cools the carburetor temperature as much as 20° less than the free-air temperature, which may be well above the freezing temperature of water. Thus, whenever the relative humidity is high, this form of icing can occur, and the engine will lose power. It will seem as though you are out of fuel, because the injection of fuel into the engine has ceased. This problem can be avoided by using the carburetor heat when you expect carburetor icing to occur.

Ground icing is the accumulation on parked aircraft when frost, sleet, frozen rain, drizzle, or snow occurs. A wise pilot will remove any ice and make certain that the plane is ice free prior to takeoff during cold weather. Another, more serious type of ice forms when an aircraft taxis through pools of water prior to takeoff during cold weather. This water may be blown by the propeller onto the control surfaces of the plane, into the wheel wells, on the brake mechanisms, flap hinges, etc., and prevent proper operation of these controls — a real hazard to the safe operation of an aircraft.

Frost is another form of icing which has long been recognized as a hazard to flying. Pilots have always been cautioned against attempting to take off with frost on the wings and the control surfaces of the plane. Thin metal airfoils are especially vulnerable surfaces on which frost can form. A heavy coat of frost will cause a 5 to 10 percent increase in the stalling speed of the aircraft, even though the amount of frost is small. In flight, frost has been known to form when a cold aircraft descends through a warm, moist layer of air. When a situation such as this occurs, and the frost is on the windows and windshield of the aircraft, the pilot must fly by instrument, since he cannot see outside the aircraft. This situation fortunately does not occur often.

A few specific points to remember concerning icing and frost conditions:

1. Before takeoff, check the weather for possible icing areas along your planned route during your preflight weather briefing. Also take the opportunity to check pilot reports, and, if possible, talk to other pilots who have flown over your planned route recently.
2. If the aircraft is not equipped with deicing or anti-icing equipment, avoid areas of known icing. Water clouds or precipitation must be visible, and the outside air temperature must be below freezing for structural ice to form.
3. Always remove ice or frost from the airfoils before attempting to take off.

4. In cold weather, avoid taxiing or taking off through water, mud, or slush, if at all possible. If you have to taxi through any of these, always make an additional preflight check to ensure freedom of the controls.

5. When climbing through an icing layer, climb at a higher air speed than normal to avoid stalling.

6. Use deicing and/or anti-icing equipment during situations of light icing; but when such equipment becomes less effective, *immediately* change altitude and/or course rapidly to take you and your plane clear of the icing areas as soon as possible.

7. If your aircraft is not equipped with a static system deicer, be alert for erroneous readings from your air-speed indicator and altimeter.

8. When flying in stratoform clouds, rime icing may be extensive horizontally. Change altitude either to a flight level with above-freezing temperatures or to altitudes where the temperature is less than $-5°C$ and preferably to temperatures of $-10°C$ or below. The altitude change also may take the flight out of the clouds.

9. Freezing rain or drizzle is often due to warm frontal overrunning. Since temperatures are usually above freezing in the warm air where rain is forming at some higher altitude, you should be able to reach these warmer layers by climbing if your plane has not taken too much ice and will climb to the desired altitude. Sometimes you may wish to descend to a lower altitude to reach warmer temperatures, but this can be risky if you fail to consider the terrain. If you decide to climb out of the icing layer, do so *quickly*, since procrastination may cause your aircraft to collect an excessive amount of ice.

10. Avoid cumuloform clouds if at all possible. Clear ice may be encountered in them at any altitude higher than the freezing level. The most rapid accumulation is usually at temperatures between $0°C$ and $-15°C$.

11. Avoid abrupt maneuvers when your aircraft is taking on ice. Since the plane has lost some of its aerodynamic efficiency, it will not function as well as it does under normal conditions.

12. When iced up, fly your landing approach with extra power.

Remember that the forecaster and the people on the ground have no way of observing actual icing conditions. The only way that a forecaster can confirm his or her forecast is by having a pilot report it. Be considerate of your fellow pilots and file a pilot report.

All cloud temperatures below freezing have an ice-forming possibility. A lesser number of smaller drops favors a slower rate of icing. The smaller water drops are found in fog and low-level stratus clouds. Light rain or drizzle are indicators of small water droplets found in clouds. The precipitation does not need to be present to indicate a population of small water droplets in the clouds. The most common type of ice found in low-level stratus clouds is light rime ice. Thick extensive clouds that produce rain, as found in nimbostratus clouds, usually have an abundance of liquid water. This type of cloud system in the winter months can present serious icing conditions for protracted flight. The most severe icing will be found at slightly above the freezing level. Occasionally, clear icing is found in stratus clouds. A note to remember is that continuous icing conditions rarely occur at more than 5,000 ft above the freezing level. Usually much closer to the freezing level is where the most dangerous icing is found. Cumuloform clouds are composed of violent updrafts and downdrafts that are favorable to the formation of large water drops. Rain from these clouds is showery in nature, and the drops are large. If the temperature is freezing or colder, the water will freeze quickly and form a sheet of clear ice. Such conditions often prove disastrous to both airplane and pilot, so flying in cumulus clouds should be avoided.

Air masses most likely to have icing conditions are those that are very moist, heavy with liquid water, and at subfreezing temperatures. Moist, maritime-tropical air after being cooled to subfreezing temperatures by frontal, orographic, or convective lifting will often cause severe icing conditions. Icing may occur during any season of the year in the United States and at other middle-latitude locations. Altogether, it occurs more frequently in the winter when the freezing level is at or near the surface. Polar regions have severe icing conditions in the spring and fall, and icing is usually more severe in mountainous regions than over flatlands. This is because mountains and their orographic lifting tend to initiate icing conditions on the windward side. These vertical motions support the formation of large drops of undercooled water. The movement of a frontal system across a mountain range will generate both orographic and frontal lifting to create extremely hazardous icing conditions for aircraft. Each mountain system has selected areas of severe icing. Experience indicates that the most severe icing occurs above the crest on the windward side of the ridge. This zone usually extends 5,000 feet above the ridge when cumuloform clouds are present, but there are times when this zone extends much higher.

IFR CONDITIONS

IFR weather can be defined as any weather condition that restricts the visibility of either the horizontal or vertical and requires the pilot to fly using instruments. So, we must classify some of the weather phenomena that bring poor visibilities and low ceilings. The type and intensity of restrictions to visibility depend on the stability of the air. Stable air that resists vertical motion is favorable for the creation and maintenance of fog, low stratus clouds, continuous precipitation, haze, smoke, and other aerosol material obscuration. These phenomena cannot readily be dispersed in stable air. On the other hand, unstable air produces vertical motion that tends to lift and dissipate fog and diffuse smoke and haze. The most common producers of visual restrictions are fog, clouds, smoke, blowing snow, dust, sand, and precipitation, but there are others. And many conditions of visual restriction sneak up on the unwary pilot.

In the western United States, in the semiarid and arid regions, you frequently take off in the early morning hours with clear skies in the summertime. Everything appears to be ideal for flying. During the later hours of the morning, you notice that a few cumulus clouds have begun to form, but this does not seem to impede VFR flying. You continue your journey, because you can still fly beween the clouds, with plenty of open sky. By early afternoon, the clear and beautiful blue sky is no longer clear. You suddenly find yourself flying above a complete overcast, and the ground is no longer visible. How do you safely descend through this cloud deck of rapidly building cumulus clouds, especially when one may be hiding a mountain, as is so common in the basin-and-range area of the western United States? This is a situation when a VFR pilot is in trouble, and when a current instrument rating is most valuable. This situation could have been avoided if you had stayed below the cloud deck, but of course you cannot always underfly the clouds in mountainous terrain; often the ground seems to rise faster than the cloud layer that hides the mountain ridges with a veil of clouds. So it is wise to be alert and not permit yourself to become one of those statistics who "continued VFR flight in IFR weather."

Another aspect of "terminal flight disease" in mountainous topography is that clouds form very rapidly over mountains and in mountainous regions. The situation is devious for the pilot who is flying in canyons that appear to be VFR with excellent visibilities and high ceiling with only a few clouds in and about. Meteorological conditions may appear harmless, but they can change abruptly, leaving you with no place to go except into a granite-cored cloud. A safety rule for canyon flying is: Make certain that you have enough visual air space to permit 180° turn at all times so you can move quickly and safely out of a danger area.

Next, on to the most common of the IFR weather producers, *fog*. Fog is defined as: "A hydrometeor consisting of a visible aggregate of minute water droplets suspended in the atmosphere near the earth's surface. According to the international definition, fog reduces visibility below 1 km (0.62 miles). Fog differs from a cloud only in that the base of a fog is at the earth's surface while clouds are above the surface. When composed of ice crystals, it is termed ice fog."* Fog is the most frequent cause of visibility of less than 3 miles, and it is one of the most persistent weather hazards encountered in aviation. Visibility can change from VFR conditions to IFR conditions of less than 1 mile in a matter of minutes because of the rapidity with which it forms. This makes flight hazardous in areas conducive to fog formation. Fog is of even greater concern and presents the greatest hazard during takeoff and landing. That is why the FAA publishes landing minimums applicable to each airport. When the local visibility and/or ceilings equal or go below those minimums, the airport will be closed to aircraft operations until the weather improves.

Numerous classifications of fog have been made. Some are simple, others detailed. For our purposes, a fairly simple classification will be used in which fog will be classified according to its mode of occurrence, as air mass or frontal fog. Within each of these general classes several subclasses can be described. In general, fog occurs under stable conditions, usually when an inversion is present with calm or light wind conditions. Fog occurrence is most common where there is an ample supply of atmospheric moisture. Conditions necessary for the formation of fog are moist air and an active cooling process. Fog is more common during the winter months than during the summer; on clear nights the atmosphere possesses a stable temperature structure with ample water vapor content. If the air is very moist, the formation of fog will occur at the ground surface under the inversion. The inversion is an expression of surface cooling.

---

*Huschke, R.E. *Glossary of Meteorology*. Boston: American Meteorological Society, 1959.

It is characteristic of air masses that were originally warmer than the surfaces over which they are now passing or resting. Some fogs owe their existence to the increase of atmospheric moisture without appreciable cooling, such as prefrontal fog situations. However, air first must be rendered stable by surface cooling.

Fog is formed by either of the following atmospheric processes: (1) air cooled to saturation, or (2) moisture added to the air until it becomes saturated. Using this information, we can formulate a few rules of thumb to assist us in anticipating the formation and clearing of fog. We find that fog frequently forms:

a. During the night or early morning when the skies are clear and the air is still, if the temperature-dew point spread is less than 15°F at dusk.
b. When the temperature-dew point spread is small and continuous light rain or drizzle is falling and other atmospheric conditions are favorable for fog formation.
c. When a decreasing temperature-dew point spread at the surface indicates the likelihood of fog. Particular caution is advisable if the spread is decreasing in magnitude to about 5°F.
d. Dense fog is considerably more likely in industrial urban areas than in rural areas.
e. Shallow fog usually clears out within a few hours after sunrise if there are no clouds above the fog.
f. Thick fogs or those with opaque cloudiness overhead thin out or clear very slowly during the day.

Fog is also classified according to the easily recognized meteorological processes by which it forms. Some of these are:

1. *Radiation fog*—Occurs mostly in the fall when the air is still moist from vegetation and radiational cooling is prevalent. Fogs of this variety are usually very shallow and burn off soon after sunrise or later during the morning. The conditions conducive for the formation of radiation fog are clear skies, little or no wind, and high relative humidity. These atmospheric conditions usually occur late at night or in the early morning when the air is stable and cool. As the insolation strikes the ground surface, it begins to warm; as the lower layers of the air begin to warm, the lapse rate changes from stable to either neutral or becomes unstable.

The height of the inversion and thickness of the mixing layer expand as the temperature increases. The mixing layer of the air under the inversion thins the fog as it occupies the enlarging volume. If the temperature exceeds the dew point temperature, the fog will evaporate into the warmer and dryer air. When clouds are overhead and retard solar heating, the visibility will improve very slowly.

2. *High inversion fog*—A winter fog with strong radiational effects accompanied by subsiding air of the winter anticyclones (high pressure). These are common in the Central Valley of California, in the Great Basin, and in intermountain valleys of the western United States and Europe.

3. *Advection-radiation*—Occurs when warm, moist air overruns cooler air from lakes and seas. It is found inland from the Gulf of Mexico and along the southern Atlantic coast of the United States. Its occurrence is common adjacent to the Great Lakes.

4. *Upslope fog*—This is found on the High Plains east of the Rockies and along the east slope of the Rockies. Winds of 20 or 30 knots are often associated with these fogs.

5. *Frontal fogs*—These fogs can be divided into subgroups, that is: (a) Prefrontal fog, which forms when the winds are weak ahead of the front; the fog can form even when the temperature is increasing as the moisture is increasing. (b) Postfrontal fog, the product of warm rain falling into the cold polar air supplying both the moisture and the cooling necessary for its formation. (c) Frontal-passage fog, resulting from the lowering of frontal clouds, especially scud clouds during a frontal passage when the relative humidity is very high; frontal fogs are usually short lived.

Smog is another source of IFR weather. This term is a combination of the words smoke and fog. The word was first used many years ago to describe the poor visibility conditions in London, England, which resulted from a mixture of black smoke and dense fog. Some smoke particles form hygroscopic condensation nuclei that will permit condensation to occur at relative humidities of less than 100 percent. The popular usage of the term smog has resulted in the application of the word to any situation of poor visibility in polluted air. Now, under very stable atmospheric conditions, a temperature inversion and a combination of smoke, fog, and other atmospheric pollutants create smog that restricts visibility at some air terminals near urban areas.

This atmospheric condition results in poor visibilities and low ceilings that are hazardous to aviation. A typical southern California *haze*, when combined with urban pollution, has become a serious annoyance to aircraft operations.

Another cause of IFR weather is stratus clouds. Like fog, they are composed of extremely small water droplets or ice crystals that are suspended in air. An example of a low ceiling is when stratus clouds hover a few hundred feet above the ground surface. If the same cloud were on the ground, it would be called fog. Often, both stratus clouds and fog are present with only a few hundred feet separating them.

Other IFR weather producers are *dust* and *haze*. These differ, in that haze particles are so small that individual particles cannot be seen and identified. Visibilities are generally sufficient to permit aircraft operations during periods of atmospheric haze. The primary danger occurs when flying the landing approach while facing the sun; the haze will restrict the pilot's vision by forming a glare that makes it impossible to see the runway. This problem can also be present for aircraft taking off. Inversions often trap haze and/or smoke and contribute to visibility restrictions. The visibility conditions generally improve as solar heating causes the inversion level to lift and increase the thickness of the mixing level during the warm afternoons. This improvement of visibility is usually much slower than for fog. Fog usually will evaporate as the air temperature increases, thus providing excellent visibility, but a haze layer will only rise as the mixing layer and temperature increase in magnitude.

Other restrictions to ceilings and visibilities occur when strong winds lift dust, causing poor flying conditions and more IFR weather. This can occur in both stable and unstable air. In stable air, dust is lifted to great heights and may spread over large areas when picked up by geostrophic winds. Visibilities can be restricted both at the surface and aloft. When the atmosphere is unstable, dust will also be raised. Once the dust is airborne it may remain suspended, restricting visibility for several hours. Dust storms often occur with the first gust of an advancing dry frontal system. Blowing sand is more localized that is blowing dust. Blowing sand is seldom lifted to heights greater than 50 ft, but when this occurs visibilities are near zero.

Precipitation also causes IFR conditions because of its restrictions to vision. Drizzle and snow are more troublesome to visibility than rain. Drizzle occurs in stable air, accompanied by fog, frequently resulting in extremely poor visibility. Rain seldom reduces visibility to less than 1 mile, except in brief heavy showers. Rain does, however, limit cockpit visibilities, especially when it streams over the windshield; it may even freeze when the temperature is below freezing. Sometimes rain causes fog to form on the inside of the windshield. Either of these can cause an impairment to vision.

Precipitation often changes low ceilings and indefinite ceilings into obscured skies. An obscure or partially obscure sky is another important aspect of IFR weather. An obscuring phenomenon is any condition that hides the clouds or sky and reduces your vertical visibility. If the sky is totally hidden by surface-based phenomena, the ceiling is reported as the vertical visibility. If clouds or parts of the sky can be seen above or through the obscuration, the condition is defined as a partial obscuration, and no ceiling is reported.

To the pilot, ceiling and vertical visibility have entirely different meanings. As seen in Figure 8-3, you will note that when a pilot is landing an airplane and the ceiling is reported as 500 ft, as soon as the aircraft is below the ceiling the slant-range visibility is excellent. If, however, the vertical visibility is reported as 500 ft, the pilot can see *only* 500 ft in any direction at an altitude of less than 500 ft. Thus, landing can be very hazardous under these conditions. Pilots should be extremely cautious while flying and landing when any impediments to vision are present.

FIGURE 8-3 Difference between ceiling and vertical visibility.

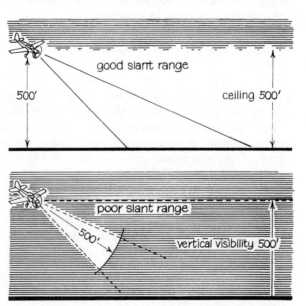

Now, let us review some of the properties of weather to assist you in becoming more alert to the meaning of VFR and IFR weather. Be especially alert for those situations conducive to the formation of fog, such as when there is a small spread between the air temperature and dew point, and especially when the skies are clear with calm winds. Another situation that improves the chances of fog formation is when moist air is flowing over a cool surface. Again, remember that fog often develops into low stratus clouds when a light wind is present to lift it. Upslope fog can form when the moist air moving up rising terrain is cooled adiabatically to the dew point temperature. Steam fog forms when cold air moves over warm water. Precipitation fog develops as rain or drizzle evaporates and recondenses as it falls through the cool air. This is most common at a warm frontal surface or behind a stationary cold front during the winter months. Low stratus clouds appear whenever a flux of low-level moist air overrides a shallow pocket of cold air that has become separated from a retreating cold air mass. Poor visibilities develop when an anticyclone (high-pressure area) stagnates and remains for several days in an industrialized urban region. Poor visibilities can also result from material blown from the surface, such as dust in an arid or semiarid region. These blowing conditions usually prevail in the spring months, but have been known to occur as blowing snow in the winter months. Southern California has blowing dust accompanying the Santa Ana winds, which also cause other problems.

As a final note, remember that if either you or your plane are not equipped to fly in IFR weather and those conditions are present, *don't fly!* During flight, if the weather ahead is worsening, just make a 180° turn. All pilots know *how*, but only the wise know *when*. This maneuver will permit you to have a safe flight, even though you may have had to detour. It is one of the safest turns you will ever make. Never think that you can go just a little further ahead into IFR weather, that it will revert to VFR conditions. It is much better to get home a day late than to return home in a pine box!

9 WEATHER AND TERRAIN FLYING

There are several locations in the United States where topography becomes an important factor in flight safety. Mountainous terrain, especially when unfamiliar, becomes an area of exaggerated concern. A general knowledge and understanding of the characteristics of atmospheric behavior in mountainous regions of the United States is important for pilots to consider. In the past, most pilots "talked terrain" prior to each and every flight along a mountainous route. This practice is still a good idea in order to obtain first-hand information from someone who has been lost there before. With the advent of better instrumented aircraft (improved engines and greater range), however, this practice is now on the wane. There are two basic ways of learning about mountain flying: (1) from your own experience, and (2) in conversation with pilots who have had much experience in flying the identical routes which you may be planning to use. Airline and industrial pilots who fly multi-engine aircraft equipped with sophisticated navigational instrumentation, radios, and radar are better prepared for almost any type of flying. Weekend pilots who fly for pleasure, or those who take to the skies for personal business, require a solid understanding of the problems of mountain flying.

Visual aspects of flight in the mountains can be deceptive. A good rule of thumb: Maintain a 2,000-ft clearance over the surface at all times. Also, pilots should rely on their navigational aids at all times so that they will not miss their checkpoints.

Temperatures and their extremes should be noted whenever flying from airfields at higher elevations. They are extremely important to flight safety, especially in the summertime. All flights using airfields at high elevations should plan takeoffs in the early morning hours when cool temperatures prevail. This type of procedure will make your flying trip to mountain areas much more enjoyable, because it avoids the density-altitude problem, along with vertical air currents and turbulence, the latter resulting from the uneven surface heating of the rough terrain. Any flight through mountainous areas should be planned along an airway, near a railroad or highway, or through a populated valley. Such a flight plan will make the trip easier to navigate, and if weather or other trouble should develop you will be much closer to assistance. A hurried route selection can lead you into a battle with unreal weather in impossible terrain! Filing a *flight plan* before you leave is imperative! In an emergency, at least the *Civil Air Patrol* will have a good idea of where to look for you. This could greatly improve your chances of flying *again*.

Here's an interesting fact: A mountain range will cut up a frontal system when it passes through. Parts and pieces of the frontal system may be found stacked to the west of mountains and ridges with associated stormy weather—all of this days after the actual frontal passage! By this time, the front itself is several hundred miles to the east of this orphaned weather.

What about takeoff from higher elevations? Does this pose special problems? Certainly. At higher elevations in the more rarefied air, a plane will require both a longer takeoff run and an extended landing roll. Also, you might expect to encounter an increased amount of turbulence during the daytime hours. This turbulence may manifest itself as cumulus or cumulonimbus buildups over the mountains. The worst time? Those warm, lazy, sunny, summer afternoons. Another difficulty of equal importance in mountainous regions is *density altitude*, the pressure-altitude adjusted for temperature deviations in the context of a standard atmosphere.

Density altitude is equivalent to that altitude in the Standard Atmosphere where density is equal to that of the air in question. Remember that it should never be used as a height reference, but rather as an index of aircraft performance for takeoff, climb, etc. It shouldn't be confused with pressure-altitude, true altitude, or absolute altitude! An approximate value of density altitude should satisfy the average light-plane pilot. High-performance jet aircraft will require more exact information, but there are charts supplied by the manufacturer of each plane that can be used. Meteorological factors are necessary to assist in density-altitude computation, especially temperature, pressure, and humidity. Pressure and humidity variations are rather minor for most cases and can be neglected, since they usually cause a deviation of just a few hundred feet in the density-altitude. The temperature contribution to the density-altitude often is vital. It causes changes of as much as 3,000 ft because the temperature may start from a morning low of 50°F and rise to an afternoon maximum of 100°F or more. Let's examine the effect of density-altitude: While flying on a hot day, the air density at 10,000 ft can be equivalent to the density at the 13,000-ft elevation of the Standard Atmosphere. This means that an airplane flying at 10,000 ft will perform as if it were flying at 13,000 ft! To make this even more confusing: When flying at the same location in the wintertime, when the air temperatures are much cooler, an airplane flying at the 10,000-ft level would perform as if it were flying at 7,000 ft in a Standard Atmosphere! You should remember that the performance of the power plant and airfoil efficiency will decrease as the density-altitude increases, and a pilot must always allow for this reduced performance of the aircraft. The length of the takeoff run and landing roll is increased, and the rate of climb is decreased. This is most critical at high-elevation airports due to normally lower atmospheric density.

The safest and most comfortable method of meeting and beating the density-altitude problem is when you are in the planning stage of your flight. If you plan to travel to higher-elevation airports you should plan to arrive and depart in the early mornings or late evenings. The chances are that the density-altitude problems that you may encounter then would be minimal. Keep in mind that the density-altitude may be as much as 3,000 to 5,000 ft higher than the true altitude. Remember also when flying in and out of high-altitude airports when the air temperatures are above normal that danger is lurking near the end of the runway: Takeoff and landing should be made using *the entire runway*. When taking off from high-elevation runways, place the aircraft in a shallow climbing attitude. Let it fly off, and never try to force it into the air. It may not go, lifting off the runway for a short distance only to fall back onto it.

Turbulence present in mountainous areas is of two varieties. The mountain has two effects on air motion. The first is *passive*, and the turbulence generated is *mechanical*. It occurs because the mountain alters the vertical and horizontal pattern of air flow about it. An example of this alteration is the mountain wave, which is found in varying sizes and strengths and is closely associated with updrafts and downdrafts. These meteorological conditions can be determined and forecast by predicting the direction and speed of the winds at ridge levels. Flying conditions of this type can be very unpleasant and very dangerous. The second variety is *active*, when a mountain develops its very own wind pattern consisting of systematic updrafts during the day which are replaced by cool downdrafts at night. A quick and easy method that can be used to check for turbulence in mountainous areas is to examine the wind field at or near the ridge level. If the winds are perpendicular to the ridges and their speeds are greater than 50 knots, severe turbulence can be expected, extending downwind 150 to 300 miles. Moderate turbulence will be present under these circumstances when the wind is between 25 and 50 knots. Extreme turbulence can be expected in mountain wave situations, in and below the level of well-developed rotor clouds. Sometimes it extends even to the ground surface on the lee side. In each of these cases, another "red flag" to watch for is the pendant cloud and telltale elliptical or lens-shaped (lenticular) clouds above and on the lee side of the mountain ridge line. This is a sign that turbulence is close at hand.

Another rule of thumb: *Never* fly up a canyon in order to clear a mountain top or ridge; low ceilings, updrafts, downdrafts, and decreased aircraft performance can be fatal. Always beware of clouds while flying in mountainous regions. They may have granite cores, and this makes penetration impossible. A pilot commonly gets into trouble when flying in unfriendly terrain by deciding to maintain VFR at the lowest reported ceiling.

Another problem in flying through and over the mountain routes is that weather-reporting stations are more widely separated than those of other areas of the country. Because of this, weather conditions encountered between mountain stations may be entirely different from those observed or predicted. Mountain aviation weather conditions greatly differ from those of the flatlands.

Still another hazard that many pilots are unaware of is the *interior valley fogs*. These fogs often form in the stable air of interior valleys. Often pilots flying from fog-free areas will find themselves overwhelmed in a fog-encased valley with overhanging low stratus added for enjoyment. This means that any airfields in the valleys will be completely out of operation. If the air temperature is about 26°F and the fog is composed of liquid water droplets, then weather modification may be successful, and the runways can be opened for short periods of time to permit limited aircraft operations. These fogs are prevalent during the late fall and winter months. Whenever you plan a cross-country flight, no matter how favorable the weather appears at your departure location, it is an excellent practice to obtain *at least* two possible routes — one alternate and one primary — when flying over mountainous terrain. This will help to eliminate the practice of going a bit further in marginal VFR weather in the hope that conditions will improve as you move ahead. If conditions begin to deteriorate, make a quick 180° turn and look for another route!

When flying a light aircraft in the high mountains, it is also a good practice not to load it to its allowable gross weight. By keeping the plane's weight below gross, you may be able to clear *that last ridge* with enough of a margin of safety. In case you find you cannot clear a ridge safely, you should always maintain enough flying room to do a 180° turn and return to the nearest airfield. Then you can remain overnight and start the flight in the early morning hours when temperatures are cooler and density-altitude is lower; this will also improve the plane's performance.

Another problem a pilot can encounter when flying in the mountains: The actual horizon is near the base of the mountains and not at the ridge line or summit of the peaks. If the ridge line is used as the horizon, the airplane may be placed in an attitude of constant climb. This and the higher density-altitude could inadvertently lead to a stall, with dire results.

At the 10,000-ft elevation, the amount and location of turbulence, along with its accompanying up- and downdrafts, are governed mostly by wind speed and direction. Ridges often present possible dangers of mechanical turbulence. Sharp ridges that are perpendicular to the wind flow will usually be the worst offenders. The wind one experiences usually varies in both speed and direction from the forecasted winds for that area. These variations are due to both the passive and active effects of the mountains. The passive effects will cause funneling and modification of the flow. Normal precautions should be taken when wind speeds reach 30 knots. Increased caution should be taken by the alert pilot when wind speeds reach 40 knots. At 50 knots, mountain waves with severe to extreme turbulence are common, and it is unwise to fly through passes or over ridges during these situations.

When the wind is blowing perpendicular to mountain ridges, and your flight takes you near the windward slopes of a mountain range, your aircraft will be lifted up and over the ridge if you are flying downwind. When flying into a headwind, you will encounter more serious problems as you approach the ridge line. To become encompassed in a downdraft is an unpleasant sensation. If altitude is insufficient, you may end up smashed against the side of a mountain. Flatland pilots prefer to fly near the surface when encountering head winds, but this strategy would be suicidal when flying in mountainous terrain. If trouble is suspected when approaching a ridge, a good maneuver is to make a slight turn and approach the ridge line at a 45° angle. If turbulence and downdrafts become dangerous, a 90° turn will then take you away from the problem area. If you were to approach a ridge head-on, you might not be able to make a turn soon enough.

When flying in a valley, you should fly on the windward slope of a mountain chain. This should provide you with additional free altitude and give you a higher air speed. If you were to fly through the valley on the downwind side, you might have to fight for altitude, especially in a light aircraft. To summarize, whenever you fly up or down a valley in mountain and canyon country, it is to your benefit to plan your flight so that it will be on the windward slope of the mountain—getting free lift. In those cases where no organized cross-mountain flow exists and all air motion in and near the mountains is produced by topography, the mountain becomes an active participant in generating its own wind system. Under these conditions the pilot will encounter updrafts on both sides of the ridge line during daytime hours. This wind pattern suggests that downdrafts will be over the valley; there they are not as dangerous as those adjacent to the mountains. This means that a light plane will tend to lose altitude over the valley, or it will rise with the rising air as it approaches the ridge on either side. The active effects of the mountains result from the insolation incident upon them, which in turn warms the surface, heats the atmosphere, and initiates convective motion. In a like manner, those mountain slopes and surfaces that do not receive solar warming (that is, the shaded side of the mountain) will possess downdrafts that may be locally intense. These should be circumnavigated to avoid problems.

During periods of VFR weather and light winds, there are many passes in the mountains that can be traversed at altitudes of less than 10,000 ft. These can become dangerous when strong winds are present, creating their own hazardous flying conditions. Pilots shouldn't use passes during periods of widespread low cloudiness, which can conceal the mountain tops and ridges.

When the air is moist and rising in the area of the updrafts over the ridges and tops, orographic clouds will appear over the mountains and along the ridge lines. Often they will be confined to narrow bands along the ridges and can be overflown in a few minutes. When cloud cover is general, instrument flight is necessary to get anywhere. The VFR pilot should leave areas of hazardous flying conditions. In an area of orographic updrafts, the pilot will often find a location of very intense icing conditions, and any flight parallel to the ridge should be avoided.

There are many preferred routes for trans-mountain flight. Many state aeronautic offices issue publications that display reliable routes. Airport facilities are available in some mountainous areas, and each pilot should investigate these before planning a flight over unfamiliar terrain. Flight Service Stations can be helpful. Valuable advice for flying these routes is available from the pilot briefer or from local pilots who are knowledgeable. When flying in areas of unfamiliar terrain never deviate from the standard route! Such a route is recommended because it is the safest, or it would not have been so designated. A small departure from the suggested route can be fatal. Electronic navigational aids often are ineffective below the ridge line or with a mountain between you and the navigational aid source. Without navigation equipment, you may find yourself hopelessly lost in rugged, hostile, unfamiliar locations with no place to go except *down*. When a pilot gets lost under such circumstances and goes down, it often takes months or even years to find a downed aircraft.

MOUNTAIN WEATHER

The types of weather found in mountainous areas are somewhat familiar to many pilots, but there are certain items of information about wind that should be reviewed. Mountains influence the wind in two ways. A passive effect is experienced by winds "passing by" mountains. Mountains modify both wind direction and speed, also generating turbulence involving powerful updrafts and downdrafts. This turbulence can be compared to water flowing over a boulder-strewn river bed. As each boulder affects the direction and speed of the water flow, so does each uneven piece of terrain affect the air flow.

In the mountains, terrain influence on flow of the wind results in vertical motions and logical flow patterns of the air over and around rugged terrain. The orientation and shape of the obstruction (that is, the mountains) with relation to the wind, the heating of slopes and mountain tops by insolation, and the steeply sloping ridges with their jagged fissures all accelerate the winds moving along them. These physical features not only generate local wind systems of the mountain-valley effect, but again reinforce the passive modification by the mountain of the winds. An interesting bit of information is that a well-defined frontal wind system exhibits fast-moving, gusting, and shifting winds. These may surge over the mountains and cascade down the slopes, often displaying unbelievable velocities of hurricane-force winds. These damaging downdrafts often abruptly change direction. Frontal passages may be accompanied by dangerous wind shear and turbulence; they are rather seasonal in nature, and are most fickle and common in the autumn and winter months.

Some of the rules of thumb for the flyer to use are:

1. Ridge level winds in excess of 20 knots indicate that a pilot should proceed with caution, and when the wind speeds perpendicular to the ridge level are greater than 30 knots, delay the flight!
2. A mountain ridge line should be crossed at an angle rather than head-on to allow the pilot to retreat from the ridge with the least amount of downdraft encountered.
3. Remember the active effects a mountain has on generating its own wind system even though the prevailing winds of the general circulation system may be westerly. Certain diurnal patterns will exhibit themselves as the winds flow up a valley during the daytime hours and down the valley at night. This is due to uneven heating from the sun. Various slopes and valleys are exposed and shaded from the sunshine during the different periods of the day.
4. The late-evening pattern is a reversal of the daytime pattern of flow. As the sun no longer heats the slopes, the higher parts of the terrain lose heat by infrared radiation at a greater rate than do the valleys. Thus, the total mountain becomes engulfed in drainage winds, with their velocities depending on the temperature differential between the mountain surface and the air.
5. Generally speaking, early-morning and late-evening flights can be made closer to the ridge line level than during the heated day because of the lower density-altitude effect.

Flight in or through mountainous regions should only be conducted when the most favorable flying conditions prevail. Such flight should only be made when the valleys along the proposed route have no hidden weather hazards. Higher altitudes are required only to clear the intervening ridges. Also, while flying in mountainous areas pilots should keep in mind that mountains act as heat sources. Heat sources increase density-altitudes and also reduce the performance characteristics of the aircraft.

It is necessary for safety's sake that pilots have an elementary knowledge of how mountains affect weather, and vice versa. Mountain weather greatly modifies frontal systems; often, mini-weather systems are localized and cast adrift from the main frontal activity by the topography. These are very difficult to forecast and account for. They are also hard to identify because of the greatly separated observing stations in the more remote, rugged regions of mountainous terrain.

Frontal systems of the mountainous western United States behave differently than those of the rest of the country. This is mainly because of the source of moisture and the orography. East of the Rockies, the source of the moisture becomes the Gulf of Mexico to the south. In the western mountains, the source of moisture is the northern Pacific Ocean, including the Gulf of Alaska. This means that the moisture is accompanied by cold northerly air, which is greatly different from the warm moisture-laden air from the Gulf of Mexico. East of the Rockies, the warm, moist air arrives ahead of the cold frontal system, and precipitation is mostly prefrontal; this means prefrontal squall lines and precipitation. To the west, the prefrontal air comes from the desert of the southwest United States and northern Mexico. In the latter case, the prefrontal air will be warm and dry. Warm fronts are ill-defined in the western United States, and will often be dry frontal systems. The cold frontal precipitation is postfrontal. Since all of these types of fronts are accompanied by northerly flow, it follows that the poorest flying weather, that is, low ceilings and reduced visibilities, will occur at and behind the frontal surface. VFR flying weather will be found along the north and west slopes of the mountains.

10 WEATHER TO SOAR

Soaring (that is, flight in a powerless aircraft) is said to be "the king of sports." What is it like to cloud-float through the heavens or glide like a bird? Silent travel through a silent world, glider flight can be an extraordinary experience. But although soaring can provide you with the freedom of the birds, it also puts you at the mercy of the weather. Can you overcome many of the hazards of the weather as they are related to sailplane flight? Let's hope so. In soaring, the whims of the atmosphere command fate even more than in powered flight. The sailplane acquires the energy needed for its flight from the atmosphere and its motions. Like the sailboat, it is wholly dependent on the motions of the atmosphere as its source of movement.

A sailplane must have auxiliary power to become airborne. This power is usually supplied by a ground tow, an aerial tow, or a catapult. Once the sailplane is airborne and the initial thrust is spent, local updrafts or thermals must be located and entered into if the flight is to continue. At this time, the free-flying sailplane becomes totally dependent on two factors: (1) the skill of the pilot, and (2) the vertical motion of the atmosphere, for lift, distance, and duration of flight. The vertical motions are classified as to their source and cause. These are either thermal updrafts or updrafts of orographic origin. They are known to the sailplane pilot as "lift" when acting on the glider. For the sailplane to maintain flying speed and avoid stalling, it must continually be in a sinking attitude. Thus, it can be said that a sailplane's success in flight is dependent on the skill of the pilot and the ability to use gravity and updrafts as the sole source of thrust in order to maintain flying speed. This is the only method by which a powerless aircraft can fly.

The necessary lift to enable the sailplane to stay aloft must come from the weather processes of the atmosphere, that is, the vertical motions of the air. These air movements are skillfully used by good sailplane pilots. So it must be obvious that the various sources of these atmospheric motions must be understood by the soaring aviator. The vertical motions, as noted, will be classified according to their origin: (1) *thermal*, and (2) *mechanical.*

"Thermals" are generated when air resting on a quasi-homogeneous surface (which absorbs insolation) rapidly heats the overlying layer of air. As heating occurs, the air expands and becomes less dense and more buoyant than the neighboring air resting on an adjacent but slower-to-warm surface. This buoyant air rises. After reaching a critical height, it separates from its thermal source, the heated surface, and rises as a buoyant bubble of air. After it leaves its initial resting place, less warm nearby air slides in over the fast-warming surface and the heating process is repeated, becoming most active during the last morning and early afternoon hours. This heating continues until the sun's rays reach a low angle of incidence or until clouds intercept the insolation and the surface is no longer an effective heat source.

Mechanically initiated updrafts of vertical motions occur whenever the horizontal air flow is confronted by a barrier the flowing air cannot penetrate. This causes the air flow to be deflected upward and over the obstruction. Thus this rising air forms an updraft on the windward side of the barrier. This obstruction can either be an orographic feature or a colder, denser air mass. When warm air advances and glides up over colder, denser air, it is referred to as a warm front. If advancing cold air slides under warm air (forcing it to rise), a cold front is present. Each of these examples demonstrate mechanical lift.

Let's consider the *mountain wave*. This shows the passive effect mountains have on air motion. As the wind blows toward the mountain from a perpendicular or nearly perpendicular angle, the air cannot penetrate the mountain. Instead, it is forced to rise over it. This climbing is what meteorologists call *orographic lift.* Orographic lift forms on the windward side of a mountain range. A compensating downdraft on the lee side will develop. On those occasions when the wind speed is in excess of 50 knots, associated updrafts and downdrafts may extend to the tropopause, and even higher into the lower stratosphere. The lee-side downdrafts may extend to the surface. Large waves often form above the crest, with a secondary wave about 100 miles downwind, and a third wave about another 100 miles downwind. Complete overturning may occur under the wave-crest clouds at lower levels. These waves are called standing waves or mountain waves. The visual manifestation of the occurrence of a mountain wave or waves is the appearance of a pendant cloud hiding the mountain top or ridge. Above the pendant cloud often can be observed a standing lenticular cloud or clouds that are named according to the heights of their bases (that is, altocumulus standing lenticular, cirrocumulus standing lenticular, etc.). These standing wave clouds form on the crest of the waves. Beneath the lowest standing lenticular clouds often is a series of rotating eddies. At low levels, overturning rotor clouds may form, but only when sufficient moisture is present or when dust or debris is entrained from the surface to make the rotor visible. The danger exists when soaring in a mountain wave near the lowest cloud. Any attempt to soar under the standing lenticular cloud can be dangerous. This is because of the rotor cloud, which may be visible, but whose rotor *motions* cannot always be seen.

THERMAL SOARING

To the sailplane pilot, the physics of *making* hot air bubbles may not be interesting, but the *finding* of thermals is exciting indeed. Let's learn how to identify the areas that are conducive to the formation of such rising bubbles of heated air. Areas that emit thermals have common features by which they can be recognized. This information could prove to be very useful when flying over unfamiliar territory. First, such areas have dry, clear sections with higher local temperatures than moist or vegetated surfaces. These dry spots are often found on raised terrain, with better solar exposure making them more likely to act as a thermal source. Moist surfaces are generally poor sources for thermal development or of convective activity. This is because most of the heat received is used to evaporate water! This implies that thermals should be sought over bare ground, fields of stubble, dry lake beds, black-topped surfaces, tilled fields, hills, mountains or urban and industrial areas. If you plan to use thermal updrafts or the urban or industrial areas as a source of lift, then make certain you have sufficient altitude to enable you to completely clear populated areas, in case you fail to find that elusive thermal needed for a successful flight. Remember, thermals are most frequent during the daylight hours, and those associated with orography are usually the most easily located.

If a circulation regime is to function, the thermal updrafts must be compensated by a subsiding air motion. This may be in the form of downdrafts or perhaps a broader downward settling of air. Thus, it is obvious that the fast-rising thermals occur over a small fraction of the local convective areas, while the gentle downward movement of air predominates over a larger area.

Since thermals are primarily creations of intense solar heating of the surface, soaring associated with them is mostly restricted to daylight hours with clear or scattered sky cover. As expected, during the nighttime hours thermals are less frequent. Then radiational cooling at the ground causes the atmosphere to become stable, often resulting in an inversion at or near the surface. When this situation occurs, vertical motion is suppressed. This situation remains until the sun rises to burn off or lift the inversion, permitting the thermals to be reestablished. Then soaring may begin anew, and the joy of motorless flight can return as an option of recreational activity. The vital ingredients for the soaring pilot's enjoyment is the skill of operating the glider while simultaneously understanding the physics of vertical motions of the atmosphere. Much of the skill required in the recognition of the sources of updrafts comes from matriculation in the "college of hard knocks."

Thermals come in a great variety of sizes and shapes. When surface heating is intense and prolonged, thermals develop in a steady column of vertical motion in a chimney shape of rising hot air. These are quite common. The sailplane aviator can find thermal lift available at most altitudes. Birds often are soaring in these blasts of hot rising air, and will assist in locating such thermal activity in the atmosphere.

Bubble-type thermals are formed when the surface heating is slow or intermittent. When the intensity and its resulting buoyancy is greater than the rate of heating, the growing bubble may be pinched off. It rises intact whenever a stable layer is reached. "Bubble formation" may be aided by shading from a cumulus cloud. A modification of the bubble-type thermal is the vortex shell. This is much like a smoke ring or a doughnut, with lift in the center and downdrafts occurring on the periphery.

"Thermal streets" occur at times when thermals appear to organize themselves. They are believed to form when wind changes are slight throughout the mixing layer, which is beneath a layer of very stable air. "Streets" are rather evenly spaced.

They may occur in clear air or with convective clouds. If convective clouds are present, the distance between streets will be two to three times the height of the cloud tops. Downdrafts are found between the streets and vary from being moderate to strong. Cumulus streets are often found behind cold fronts (in the cold-air sector) in which flat cumulus-type clouds develop. Because of the thermal activity associated with thermal streets, an aviator can soar beneath the cloud street and maintain altitude without circling. This makes flying a bit easier.

Another source of lift can be found in the thermally generated local wind patterns of an orographic setting. These follow a diurnal pattern, with the air motion being upslope and up-valley during the daylight hours and reversing direction between sunset and sunrise. The daytime upslope winds have a greater magnitude than do the downslope drainage winds. Defant (1949) analyzed the double-theodolite pilot-balloon ascents of Riedel on the slopes of the mountains near Innsbruck, Austria. This was done to determine the wind components parallel to the slope as a function of the height above the slope. The investigation demonstrated that maximum wind speeds of more than 12 ft per second were observed at a height of 20 to 40 m above the slope surface. To a sailplane pilot, in order to use this source of lift, this means flying with the aircraft wings dangerously close to the surface of the slope if the sailplane has a wingspread of 16 to 18 meters. Such a source of lift has been used by German sailplane pilots to cover distances of up to 30 km, using the upslope wind from a single hill, *without* losing altitude. Upslope winds of these magnitudes have been measured during the winter months on mountain slopes where ski jumps have been erected. These studies have shown that the faster the upslope wind speed, the greater the distances better skiers made on their jumps. Glider pilots can also take advantage of vertical winds, especially over the top of hills or mountains. These sources of lift are more easily distinguished than many of the flatland sources. The upslope winds are generally faster than the downslope winds because of their associated temperature gradients, which are stronger during the daytime. These horizontal temperature gradients develop because of insolation absorbed on the slopes, which warms the air much more rapidly than the free air is warmed at the same altitude over the valley.

All pilots scan the skies for convective activity. Aviators of powerless aircraft must utilize updrafts if their flight is to be sustained. Vertical motions create turbulence for powered aircraft, so many pilots of powered aircraft avoid areas that are conducive to thermal formation. Thus, for different reasons, the pilots of both powered and powerless aircraft will always be alert for the signs of convective activity, one avoiding thermals, the other using them. In fact, the soaring aviator will devote a large percentage of flying time chasing after the local convective cells as the primary source of lift. Peter Dixon estimated in 1970 that 80 percent of all soaring activity depends on thermal lift.

Dust devils are another form of thermal activity. The temperature gradient near the ground-air interface becomes very stable during the hours around noon. The upward swirling masses of air initiate rotational motion and a microvortex is formed in either a clockwise or counterclockwise direction. Studies of dust devils have shown that their direction of rotation appears to be decided by chance. In 1936, Flowers reported observing 375 dust devils; of these, 175 were rotating in a clockwise (anticyclonic) direction and 200 rotated in the opposite direction (cyclonic). Ives observed in 1947 that a dust devil which came almost to an end was able to revive itself when it came in contact with an obstruction and began rotating in an opposite direction. Dust devils and sand devils have their origins in the layer of air close to the ground which is characterized as the boundary layer on the atmosphere, in which eddy exchange increases rapidly with height. The temperature gradients near the air-ground interface during the middle of the day become very unstable. Horizontal temperature contrasts become very important. These thermal instabilities and the overturning of the superheated layers of air in the boundary layer are an exceptional occurrence, accompanied by dust devils and sand devils. The upward swirling of the air entrains dust, sand, and other debris from the surface, making the dust whirl visible. As it slowly wanders, it engulfs other superheated layers of air to maintain itself. In arid regions, dust devils are regular features of hot afternoons. Dust devils are thought to be initiated by irregular features in the terrain, like a passing automobile, a gust of wind, or a deflating of a rubber raft.

On one occasion during the month of August 1962, while working at a research station on the southern end of the Great Salt Lake, the author sighted a series of dust devils. These were initiated as the wind moved up the Tooele Valley and curved eastward along the protruding edge of the northwestern tip of the Qquirrh Mountains. Dust devils were created at 40-second intervals as the wind blew around the northern edge of the mountains. They seemed to travel in a northeastward direction across a black-topped highway, a railroad track, and a beach, only to dissipate as the lake was finally reached. The parade of dust devils was traveling in a line, with three or four passing along the path toward the lake at any given time. A lake destination meant quick suicide for the "little devils." It was a very unusual sight! My observation point was half a mile east of the line of dust devils. The family of swirling dust occurred during a time of severe drought. Very hot, dry weather prevailed, and the lake level was at its lowest elevation in recorded history.

Such phenomena have potential as a source of lift by sailplane pilots. But care must be taken whenever dust devils are used as a source of lift. These thermal motions are very strong and are often accompanied by turbulent surrounding areas of little or neutral lift, and maybe even an area of downdraft. If approaching a dust devil, always stay above the minimum level where sufficient height is available for recovery and maintenance of flight. As a rule of thumb, always approach the vortex at an altitude of greater than 500 feet. Remember, too, that the dust devil's direction of rotation is by chance. So closely examine the vortex by observing the whirl of dust and debris near the surface. Also, keep in mind that the vortex column that identifies the phenomena is usually very narrow. This fact can make it difficult to keep the sailplane in the vortex.

Special care must be taken not to enter such a whirlwind in the direction of rotation! This means you will sustain a tailwind, which will cause the sailplane to lose airspeed and altitude while increasing the chances of stalling. Such a mode often results in an increased ground speed and a high-speed long turn that throws the plane out of the vortex wind because of the added centrifugal force resulting from the summation of the glider's air speed and the tangential speed of the vortex. Such increased linear speed makes the turning radius too great for the plane to remain in the whirlwind, and out it goes! On the other hand, if the dust devil is entered counter to the direction of rotation, the airspeed should be reduced to a minimum consistent with the minimum flying speed of the glider. The inertia of the glider thrust causes a surge of air speed, which means it could easily exceed its "redline."

When flying in and around dust devils, you should be aware that the core of the vortex has little if any lift and could subject the plane to severe to extreme wind-shear turbulence. This, of course, would precipitate a loss of air speed and stalling, a horrible thought. Thus, always avoid flying into the core of the vortex. Such a practice is not conducive to the safe flight of sailplanes. Stay alert and don't let your thoughts wander.

When convective clouds form, this is a sign that thermal soaring is great! Fluffy, white cumulus clouds are signposts leading the way to otherwise elusive thermals. We have learned that rising air cools as it expands, and if it possesses enough moisture the cumuli clouds will start to form. These clouds are "vendors" of thermal activity. Another point to remember is that single convective cells (thermals) grow and die. A cloud advances through stages: growing, maturing, and, finally, dissipating. This means that if you plan on using convective clouds as a source of lift, you should watch the various stages of the cloud life cycle and then use the cumulus growing stage for flight. The reason for this is to benefit from the maximum lift these clouds can generate. A person usually will find good thermal activity beneath the growing cumuli. These growing clouds are usually small.

As the cumulus clouds become larger and intercept a greater amount of sunshine, they reduce the heating of the thermal that created them. This situation is part of a self-destruction mechanism built into various cloud systems in order to equalize atmospheric energies so that they won't become too concentrated at any one location. Of course, whenever a cumuloform cloud begins precipitating, it initiates its own destruction, murdering its energy of creation. (The same can be said of some aviators, judging by the actions they take.) In this particular case, the self-destruction begins when the shade from the cloud reduces the effectiveness of the surface thermal, cutting off its heat supply and temporarily arresting the thermal source until the cloud vanishes or drifts off. Usually, the wind permits insolation to rewarm the surface and repeat the process. This is one of the processes responsible for intermittent rising of warm air bubbles.

Another point: As cloud cover increases, ground source thermals will decrease. This means that when the sky is overcast, surface thermal activity decreases. Also, if the temperature structure is unstable, cumulus clouds will continue to grow and *cause* the sky to become overcast. Obviously, then, soaring between the clouds and the surface can be dangerous. So, in effect, soaring must be terminated for that day. Abundant convective activity with its accompanying cloud cover reduces thermal activity. If the sky cover is scattered, however, soaring is still possible.

A very unstable atmosphere sometimes produces towering cumulus and cumulonimbus clouds. These clouds obtain their energy for growth from the great amounts of latent heat released by the condensation process. Resulting from high-velocity updrafts and downdrafts, they often produce heavy rain, hail, and icing. Clouds of this magnitude should be avoided by a glider except if flown by an experienced pilot. Hail can batter a little glider to pieces; this is expensive, and occasionally terminal!

The appearance and most likely locations of thermals have been noted, but have we discussed the particular meteorological variables that determine their likelihood? Thermals are the product of an unstable atmosphere. Their physical dimensions and strength depend on the depth of the unstable layer they are in. The concept of instability was discussed in Chapter 6. The meteorological conditions favorable for soaring will be specifically of interest to the sailplane pilot. Certain questions come to mind, that is, the forecasting of the following to occur:

1. The earliest time soaring can begin.
2. The strength of the thermals.
3. The height the thermals will attain.
4. Cloud amounts, both convective and higher cloudiness.
5. Visibilities, both at the surface and aloft.
6. Probability of precipitation.
7. Winds, both at the surface and aloft.
8. Thermal index (TI).
9. Maximum temperature forecast.
10. Height of the mixing layer.

The meteorologist will forecast many of these parameters from the morning upper air sounding that is plotted on a "pseudo-adiabatic" chart. A soaring pilot should study and learn to use the pseudo-adiabatic chart. It greatly assists in determining the thermal index of the atmosphere. The National Weather Service uses the *Stuve diagram* and the Air Weather Service (USAF) uses the *Skew T log P diagram.*

There are essentially five lines of importance on each of the thermodynamic diagrams. Each describes a definite parameter of the atmosphere. These are fully explained in the Appendix. Sufficient information is available for most general-aviation and sailplane use if data are used only up to the 400-mb pressure surface. Besides plotting the mandatory information, the significant level data should be plotted, with lines connecting temperature data from the ground surface to the top of the sounding. This procedure is repeated for dew point temperatures.

Atmospheric tendencies for vertical motion can be ascertained. We know that if the displaced parcel of air is warmer than its new environment, the air will rise and cool dry adiabatically and will be unstable. This rising air is called thermals by sailplane pilots. Thus, vertical temperature features assist in determining the possible vertical motion that can be used for lift by the sailplane pilot. This makes the criteria for atmospheric stability an important tool. Instead of comparing the environmental lapse rate with the dry adiabatic lapse rate, we can compare it with the moist adiabatic lapse rate. This lapse rate (which cools at a lesser rate than does the dry adiabatic lapse) means that instability is more likely to occur when air is vertically displaced in a saturated environment. These are the sources of the convective clouds and convective cloud motions of the atmosphere. Again, a source of lift for the experienced aviator of motorless aircraft.

A morning sounding also is a good tool to use to forecast maximum temperature expected to occur during the afternoon. This can be done by examining the sounding and the depth of the mixing height estimated to occur for that day. After estimating the projected mixing height, the dry adiabatic slope that intersects the morning sounding at the mixing height is traced to the surface of the earth, and the surface temperature of the air (which is warmed dry adiabatically) is followed to the surface and is the expected maximum temperature that will occur during the afternoon.

Thermals depend on the heating of the surface air. The air expands, and being less dense than its environment, also rises. As it goes up, cooler air replaces it. Its upward strength depends on the temperature difference between it and its environment. If we think of a total picture (of vertical motions of the atmosphere), a temperature difference or thermal gradient between the subsiding and rising air exists. A rule of thumb is: The stronger the gradient, the stronger the thermals. In order to use this principle and approximate the temperature difference, the Thermal Index (TI) is determined for any level in the atmosphere. The 850-mb and 700-mb levels are most widely used, because the 850-mb level is located at about 4,700 to 5,000 ft and the 750-mb level is around 10,000 ft. These altitudes are in the domain of most sailplane operations, that is, routine soaring, and the temperatures are available because these are *mandatory levels* for atmospheric soundings.

The TI requires that three temperatures be used in determining its value. These are the surface-maximum and the 850-mb and 700-mb temperatures. The surface-maximum temperature is then lifted dry adiabatically to the 850-mb level. The dry adiabatic temperature of the lofted air is then subtracted from the measured 850-mb temperature. This becomes the TI for the 850-mb level. (The 700-mb temperature is used for stations at higher elevations.) If the difference is negative, lift is predicted. The greater the negative magnitude of the TI, the stronger the lift or vertical motion of the air! Large negative values of -8 to -10 or greater suggest very good lift and greater than normal periods of soaring for the day. Thermal indices of high negative value are strong and can be maintained even on a windy day. A TI of -3 indicates that soaring is possible, but indices of -2 to 0 mean that thermal soaring is marginal. Positive values further diminish the chances of using thermals as the source of lift.

As with all forecasts, a miss of the maximum temperature, with its resulting erroneous thermal index, will greatly modify all expectations of soaring conditions. This evaluation of lift or updrafts can actually indicate the maximum height the thermals may attain. This height will be indicated as the intersection of the dry adiabatic slope that passes through the maximum temperature and the environmental temperature (the plotted sounding). The height of this intersection is the mixing height of the surface air and the upper limit of vertical motions and soaring. After the vertically moving air becomes saturated, and prior to reaching the dry adiabatic intersection with the sounding lapse rate, then its adiabatic cooling rate will change from the dry adiabatic lapse rate to the lesser moist adiabatic lapse rate. The air now becomes convectively unstable, and the new maximum height, if it exists, will be where the moist adiabatic lapse rate intersects the environmental sounding. This may not always mean excellent soaring at greater altitudes. The moist adiabatic lapse rate is found in clouds usually cumuloform in nature, with all of their attendant problems and dangers.

Often the National Weather Service will have no upper-air sounding near the soaring site. They do, however, usually provide a maximum temperature forecast for such locations of interest. This means that determining the height of maximum lift is difficult, if not impossible. A utilization of the nearest upwind rawinsonde report can be obtained by telephone (long distance); this can then be plotted and analyzed to give a "ballpark figure." Upper-air soundings are representative of a rather wide area.

If sailplanes are to be launched with an airplane tow, then the tow plane could make a vertical temperature sounding. This sounding should measure the temperature both on the ascent and descent of the sounding flight. The temperature for each elevation will differ between the ascent and descent. The variation of readings between the two modes is caused by several factors, including the time constant of the thermometers, so these should be averaged for each height. When these readings are plotted on a pseudo-adiabatic chart, the TI can be determined. The sounding flight should read the thermometer at thousand-foot intervals, which is much easier because the altimeter measures height in feet.

Cross-country soaring in mountainous regions utilizes the orographic lift formed by the passive action of mountains on air flow and the thermal lift of daytime heating of mountain ridges and slopes. These sources of lift provide the soaring aviator with excellent opportunities to soar for great distances, especially during the daylight hours of favorable solar heating. Once away from the mountains, a soaring sailplane pilot must look for another source of lift. If a cold front happens to be in the neighborhood, excellent cross-country soaring is found behind the front in the cold air. This type of thermal soaring is found in ideal patterns, and four factors contribute to the soaring possibilities:

1. The cold air is usually dry, and thermals build to relatively high altitudes.
2. The polar air is colder than the ground; thus, the warm ground with the aid of solar radiation is warmed even further. This in turn warms the atmosphere, and thus the warm ground with the aid of solar radiation is warmed even further. This in turn warms the atmosphere, and thus thermals are initiated rather early in the morning and may last until the late evening. Sometimes soaring is possible even after sunset.
3. Frequently, cold air aloft at high altitudes moves over the low-level air. This serves to intensify the instability and may strengthen intensity and frequency of the thermals.
4. The wind profile often favors thermal *streeting*, a real boon to speed and distance.

These same four factors are associated with cold-front passages over mountainous regions in the western United States. The mountains break up circulation patterns of the cyclone as it moves eastward over them. The western mountain regions have a unique advantage, as the air is predominantly dry. This dryness permits much daytime thermal activity plus an orographic contribution (both in the passive and active sense of the word). Such conditions greatly favor cross-country soaring activities for both long and short distances, with best conditions present during the daylight hours. The best location for distance soaring is along the

High Plains corridor adjacent to the Rocky Mountains, reaching from Canada to Mexico. A great many records have been set at the national and international meets held in western Texas. The ground cover ranges from bare to short grass. Such features favor strong thermal development of vertical air motions. The prevailing winds blow from the south and are moderately strong, providing additional energy for the northbound cross-country aviator.

Yes, using frontal lift as the lifting mechanism is another source of power for soaring and powerless flight. This type of lift, however, is very transitory and can be dangerous. Also, seldom is the front oriented in the direction you wish to go! A dry front may be a better source of lift for soaring, since the convective clouds are not as evident; but this also can be somewhat of a problem, since dry fronts are often difficult to see. If a front is known to be in the vicinity, you should recognize the lift associated with the front by the sailplane's reaction. Keeping in the lift on the leading edge of the frontal activity can be a problem. Often this is accomplished in the same manner as soaring in desert regions while using dust devils as a source of altitude. This means watching the ground conditions for indicators of frontal activity, that is, ground gusts and dust blowing at the leading edge of the frontal activity.

A common local circulation system that occurs along the coastal regions is the sea breeze during the warm seasons. This pleasant breeze from the sea is diurnal in nature. It begins in the forenoon, reaches a maximum during the late afternoon at the time of greatest solar heating, and reverses itself after the sun has set and the land surface has cooled to a temperature equal to the water temperature. As the land surface continues to cool (to temperatures several degrees cooler than the slower cooling water), the wind reverses itself and blows from the land surface to the water. This happens during the evening and nighttime hours. During the daytime period, the leading edge of the advancing cool air of the sea breeze moves inland and converges over the land surface. The converging air rises, and often a line of coastal cumulus clouds form inland, with many similarities to that of a weak frontal system. This line of convergence with its associated rising air is an excellent source of lift for the coastal sailplane pilot. The transition zone forming between the cool, moist air of the sea and the warm, dry air inland is often shallow and a narrow pseudo-cold front. The distance penetrated inland by the cool, moist marine air is dependent on the horizontal temperature gradient of the land-sea surface, the general wind flow, moisture, and terrain.

In regions of marine climates, often a cool upwelling water makes it favorable for the development of a strong sea-land temperature gradient with an accompanying sea breeze system. Since these are of rather a local nature, a well-developed cyclonic system can easily overwhelm these systems. Therefore, a sea breeze is most likely when the ambient pressure gradient is weak and the wind light. When the layer of advected air of the sea breeze is deep, the frontal activity initiated by the sea breeze has been known to trigger cumulonimbus clouds when the lifted air contains enough moisture. Most of the time, the cumulus clouds are of limited vertical extent. As one would expect over vegetated surfaces where the air is moist, the sea breeze cumuli are the rule; in arid regions, few or no cumulus clouds develop with the sea breeze front.

Coastal mountain ranges or even irregular or rough terrain often amplify the sea breeze front. Such terrain also may generate sea breezes originating from many different locations. The active effect of mountains often increases the sea breeze because each is mainly a local thermal circulation pattern. Thus, a more intense wind system is likely to occur when these two types of thermal patterns are combined. The physics of the two are the same. Sea breezes have been known to extend to the lee side of low-lying hills and mountains. When these two circulation patterns are combined, the wind speeds will intensify. The sea breezes have been known to penetrate 150 miles inland in the tropics and may even reach 50 miles in middle latitudes, and wind speeds of 15 to 25 knots are not infrequent.

As a note of appreciation, sailplane pilots have made significant contributions to aviation. They have aided the understanding of many atmospheric processes: thunderstorms, mountain waves, local circulation patterns. These contributions have made flight safer for *all* who fly.

Nowadays, other denizens of the sky are concerned with micrometeorology. But to a large extent, "flight-for-fun" enthusiasts such as radio-control modelers, parachutists, balloonists, and hang glider hobbyists employ for their use similar principles to those already stressed for sailplane pilots. There are, however, some special points worth mentioning. An extremely careful observance of natural indicators is essential, especially if high-wind flight is attempted. For hang gliding, special attention should be given to the details of flying conditions in the vicinity of obstructions. This is especially important because hang gliding often takes place in the friction layer (an area particularly affected by mechanical turbulence). Flight downwind of a large hill or ridge may be considered a prime example. A rule of thumb for safety's sake is indicated here: Maintain a minimum altitude of seven times the height of the hill, except in calm or light wind conditions. Gusts due to buildings and trees are capable of unleashing bursts of energy hundreds of yards downwind, and all the more dangerous when they are unexpected. A smart hang glider pilot won't land in the wake of even small obstacles, unless an emergency exists.

Every type of aircraft has its own peculiar aerodynamic forces acting upon it. Factors such as the location of the center of gravity and the relative wind are always important.

11 AVIATION WEATHER ASSISTANCE

HISTORICAL BACKGROUND

The National Weather Service (NWS) began providing weather forecasts for aviation when the Wright brothers planned powered flight at the turn of the century. The NWS (then called the U.S. Weather Bureau) first studied and forecast surface winds at Kitty Hawk, North Carolina, in response to a request from the brothers prior to that memorable day in December 1903 when man's dream of ascent into the atmosphere became a reality. Thus, aviation and weather have been working and growing together from the beginning of flight. World War I saw the first quantum step in avaiation meteorology when Norwegian meteorologists announced development of the Polar Frontal Theory.

    The U.S. Post Office was a pioneer agency in establishing the nation's airways during the period from 1918 to 1926. Also, the Weather Bureau worked hard during this period to provide a needed service to aviation by issuing special aviation weather bulletins. In 1920, aviation forecast centers were established at Washington, D.C., Chicago, and San Francisco. By 1921, the U.S. Post Office had a chain of weather-transmitting radio stations at its airmail bases. It was also during this period that the first night airmail flight was made from North Platt, Nebraska, to Chicago, Illinois. The flight was guided by bonfires and automobile road maps. In 1924, a lighted airway was established from Cheyenne,

Wyoming, to Chicago, Illinois, and the Weather Bureau added night forecasting to its operations. The airways weather service was authorized in the same legislation that established the Civil Aeronautics Administration (CAA), now the Federal Aviation Administration (FAA). This legislation permitted the Post Office to contract with commercial airlines to carry the mail. Consequently, the Post Office bowed out of active participation in aviation.

At about this time, meteorologists discovered that in order to provide the needed support for aviation operations, better upper-air data were required than could be obtained with kites. Beginning in 1931, aircraft were used to make upper-air samplings of temperature, humidity, and pressure. These soundings were expanded from four stations in 1931 to 30 stations by 1937. During the expansion period, 12 pilots lost their lives attempting to obtain valid information in diverse weather situations. Commencing in 1938, upper-air measurements of temperature, moisture, and pressure were made with the radiosonde. This new system had the advantage of being an all-weather operation, providing upper-air information when it was most needed and most often unavailable from aircraft operations. During this same period, forecasters applied air-mass and frontal analysis to improve their forecasting products. At the end of World War II, 26 flight-advisory weather stations were established by the Weather Bureau (now the NWS) to work in close harmony with the Air Route Traffic Control Centers (ARTCC) of the CAA (now the FAA).

Improved weather forecasts were made possible with the advent of the facsimile machine in 1948. This development coincides with the increase in speed and range of aircraft. Radar also become operational as a new tool for remote sensing of the weather. Then, in 1959, the Weather Bureau established seven high-altitude forecast centers to provide improved weather service for the higher-and faster-flying aircraft.

Aviation operations in the United States are estimated to involve more than 100,000 persons in the nation's airspace at any one time. Safety has become very important, and weather is probably the most significant aspect of air safety. Air carriers now have approached an all-weather operational goal, but general aviation has not yet achieved that luxury. The general-aviation category includes business, pleasure, and instructional flying. Most of this type of flying in the United States, perhaps 75 percent, takes place under visual flight conditions at altitudes between

3,000 and 5,000 ft above ground surface. Flight along the civil airways averages about 130 knots, and most flights occur between the hours of 8:00 A.M. and 6:00 P.M. local standard time.

The NWS is the primary organization providing a weather service for aviation. Other nations have similar organizations to provide this type of service. These organizations cooperate with each other through the World Meteorological Organization of the United Nations, which standardizes observational and forecast procedures and provides an international exchange of weather information. The NWS has found it both economical and efficient to obtain the cooperation of other governmental agencies, both military and civilian, as well as private individuals and organizations. An overwhelming demand for weather service requirements of a rapidly expanding aviation community must be satisfied.

The NWS has created many processing centers. These include the National Meteorological Center, the Severe Local Storm Forecast Center, the National Hurricane Center, the High-Altitude Forecast Centers, and the National Weather Satellite Center. The aviation weather service did not begin as a planned operation but grew as aviation grew, beginning with the first flight at Kitty Hawk. A very close relationship exists between the NWS and the FAA. The NWS maintains a quality-control program for the Flight Service Station (FSS) pilot briefers. The briefers are spot-checked for proficiency in face-to-face, telephone, and radio presentations. Pilot briefers provide an important safety function for aircraft operations. They interpret weather information received by high-speed electronic communication systems and give concise oral summaries in logical sequence concerning observed and forecast weather conditions.

When a briefing is requested, whether face-to-face or by telephone, faster service can be obtained by providing the briefer with the following information:

1. That you are the pilot and not just a passenger who is interested in the weather. (You will probably also be asked to provide either your name and address or an aircraft identification number.)
2. The type of aircraft that you are planning to fly (light single-engine, high-performance multi-engine, jet, glider, or hang glider), since each presents a different briefing problem.
3. Your destination.
4. Your estimated departure time.

5. Whether you are IFR-qualified or not, and if so, whether you wish to fly under IFR conditions.

As a pilot, you need to obtain a complete weather briefing. After you have given the briefer the above information and any other information requested, you will need to obtain the following information from the briefer:

1. Weather synopsis (position of high- and low-pressure areas, front, ridges, etc.).
2. Current weather conditions.
3. Forecast weather conditions.
4. Alternate routes, with current and forecast weather conditions.
5. Hazardous weather conditions.
6. Forecast winds aloft.
7. Lift or soaring information.

If you don't obtain each aspect of information, or if you don't understand completely what information has been given, ask questions! Remember, it's your life that's on the line, not the briefer's. If the latest meteorological information is not available for a flight beyond the local area (or even for an extended local flight), you should see the following as a minimum:

1. The surface weather map and the surface prognostic charts for present and forecast positions of highs, lows, fronts, and squall lines.
2. The weather depiction chart to see the general areas where ceiling and visibilities have been poor. Relate these to major features of the surface weather map.
3. The radar summary chart for a display of weather hazards. This chart locates and describes intensities of thunderstorms and includes the current severe weather forecast.
4. Winds aloft charts along with the 850-mb, 700-mb, and 500-mb constant-pressure charts to obtain information about the current wind patterns for the altitudes of concern.
5. Latest aviation weather reports, radar observations, and pilot reports for your route, including the current weather conditions.

6. Area forecasts covering your proposed route and alternate routes.
7. Terminal forecasts for destination and alternates.
8. Winds aloft forecasts for the expected wind conditions along your route and alternate routes at your proposed altitudes.
9. Any in-flight weather advisories in effect for the routes and altitudes of concern.

With this information, you can select the best altitudes and routes to fly.

For many years, weather charts have been rather confusing to the general public; but with the advent of television weather presentations using both surface and upper-air weather maps, the public has become more familiar with these charts and their contents. The weather depiction chart was developed to provide the pilot with a graphic display of flying conditions across the contiguous 48 states and southern Canada. The plotted data on this chart include precipitation and/or other weather phenomena, sky coverage, visibilities of 6 miles or less, and cloud bases in hundreds of feet above the ground. Solid lines (IFR) enclose areas where ceilings are below 1,000 feet and/or visibilities are below 3 miles; these areas are usually shaded in red. Scalloped lines (MVFR) enclose areas where ceilings are less than 5,000 feet but greater than 1,000 feet and/or visibilities are less than 6 miles but greater than 3 miles; these are usually shaded in blue, locally. The chart is designed primarily as a tool to alert aviation interests to the location of critical and near critical VFR and IFR minimums at terminals in the United States and surrounding land areas. It should be recalled that weather shown on the weather depiction chart is at least 90 minutes old by the time it is displayed. This suggests that the pilot should consult the latest aviation weather reports in order to update the weather depiction chart information. The weather depiction chart is designed to give the pilot a bird's-eye view of the general weather situation. It is used most effectively with the surface weather chart, which manifests the causes of the weather indicated on the depiction chart.

Radar summary charts are prepared from radar observations made by the NWS and FAA in the United States. East of the Rockies, echo outlines are analyzed, and adjacent smaller areas of similar echoes are sometimes grouped as a single area. The analysis is then joined with the western composite chart. Appropriate Air Weather Service radar reports are added in. The intent of the radar analysis is to depict the mesoscale and general microscale distribution of precipitation patterns. Radar reports provide additional information about the weather phenonema within range of the radar system. Some of the information obtained in this manner cannot be observed by other means. For example, the three-dimensional structure of precipitation can only be estimated with radar. Weather radar is used primarily for detecting and tracking severe local storms, such as thunderstorms, tornadoes, and hurricanes. The storm itself is identified by the reflection from water drops or ice particles that produce the echoes on the radar scope. These echoes must be correlated with surface weather reports, pilot reports, and many other types of meteorological data to obtain meaningful radar summaries. Echoes classified as heavy or very heavy in intensity usually indicate a storm with severe or extreme turbulence, hail, and severe icing conditions. Echoes of very light intensity may indicate snow, light rain, or drizzle. If a severe weather warning is in effect for any area of the "lower 48" states, the area is outlined with dotted black lines on the radar summary chart. The identification numbers of the weather warnings and their valid times are shown in the lower left-hand corner of the chart.

Low-level significant weather prognosis charts show fronts depicted according to the symbols shown in Figures 5-2 through 5-4. Isobars are drawn as solid lines at 8-mb intervals, with dashed lines at 4-mb intervals added to define the pressure gradient. Pressure areas are encircled with dots and labeled H and L for high- and low-pressure areas. The instantaneous direction of motion is indicated by an arrow with the speed in knots. Areas of precipitation are depicted with a solid line. Broken low clouds with bases at or below 5,000 feet are enclosed by a small scalloped line. Extensive areas of fog or low stratus-type clouds are depicted by hatching without enclosing the boundaries. The freezing level is depicted by a heavy solid line for surfaces intersecting the freezing level. Turbulence of moderate or greater intensity is indicated by turbulence symbols. Thunderstorms always imply moderate or heavy turbulence.

SIGMET advisories concern weather of particular significance to the safety of transport aircraft (large multi-engine) as well as smaller aircraft. They are issued whenever any of the following conditions are known to exist or are expected to begin within 2 hours: tornadoes, lines of thunderstorms (squall lines), large hail (¾ inch or more in diameter), severe or extreme turbulence, heavy icing, widespread sand or dust storms that reduce visibility to 2 miles or less.

AIRMET advisories concern weather that is potentially hazardous to small, single-engine and twin-engine aircraft, but not necessarily hazardous to large transport aircraft. They may also be suitably applied to gliders and hang gliders. They are issued whenever the following are known to exist or are expected to occur within 2 hours: moderate icing, moderate turbulence, winds of 40 knots or greater within 2,000 feet of the ground, the initial onset of visibility of less than 2 miles or ceilings below 1,000 feet, poor ceiling and visibility conditions near mountain ridges and passes.

For a nonstop trip, the weather briefing should be as accurate as possible, and will be more complete if the pilot provides the briefer with some advance notice. The best weather briefings generally are conducted face-to-face between the pilot and briefer. This personalized service continues as an important part of the presentation. In recent years, the phenomenal growth of the number of general-aviation pilots has required other methods of distributing weather information. To meet this rapid growth, the NWS and the FAA have located about 600 briefing offices at busy airports. Since there are over 8,000 airports in the country, it is obvious that some pilots cannot take advantage of a face-to-face briefing. Pilots departing from an airport with neither a Weather Observing Station or a Flight Service Station may call the FSS on a telephone service provided by the FAA at no cost to the pilot. The numbers to call the weather briefing service at certain busy locations are available only to pilots and are listed only in the pilots' information manual, not in a local telephone book.

In spite of the various weather briefing methods available, many general-aviation pilots receive an incomplete weather briefing or none at all prior to takeoff. This lack of weather knowledge and failure to obtain a weather briefing undoubtedly accounts for a considerable number of accidents. In a recent year, there were 4,400 mishaps among pilots in this category; 420 involved a fatality. Among the fatal incidents, 33 percent were caused by loss of control of the aircraft by a pilot who was not instrument-rated flying in adverse or IFR weather conditions. The lack of availability of adequate weather services accounts for only a small percentage of aircraft accidents. There are few airstrips that do not have free telephone service to the nearest FSS to obtain a briefing. This is a very inexpensive insurance policy for a safe flight, yet many pilots do not avail themselves of this service!

Low ceilings and poor visibilities contribute to a large number of weather-connected accidents. Strong and changeable surface winds also cause a considerable number. Other causes of weather-related accidents are in-flight turbulence, snow, hail, structural and power-plant icing, thunderstorms, vertical air motions, and smoke haze. Some weather-related accidents could be avoided if pilots had a better understanding of the weather and participated in good weather briefings prior to takeoff. Obviously, some briefings should tell an experienced pilot to *cancel* the flight! Pilots who fly without a weather briefing offer some rather flimsy excuses. The most common are that it is "inconvenient to obtain," or "the weather did not appear to be bad," or "the flight service station was inaccessible." Nonsense! If you are flying out of a small airfield that does not have a FSS or NWS office nearby, you can always telephone the nearest FSS to obtain a telephone briefing. If the airfield doesn't have a toll-free telephone for obtaining a briefing, you may be able to tune your radio receiver to the continuously transcribed weather broadcast. Still, some pilots apparently have not yet learned to respect the weather.

Recent studies of aviation weather forecasts have revealed the following:

1. Up to 12 hours and even beyond, a forecast of good weather (ceiling of 3,000 feet or more and visibilities of 3 miles or greater) is more likely to be correct than a forecast of conditions of poor flying weather (ceiling below 1,000 feet or visibilities of less than 1 mile).
2. However, for 3 to 4 hours in advance, the probability that conditions below VFR minimums will occur is more than 80 percent, if such conditions are forecast.
3. Forecasts of a single reportable value of ceiling or visibility instead of a range of values implies an accuracy that present forecasting systems do not possess beyond the first 2 or 3 hours of the forecast period.
4. Forecast of poor flying conditions during the first few hours of the forecast period are most reliable when there is a distinct weather system such as a front, a trough, precipitation, etc., which can be tracked and forecast, although there is a general tendency to forecast not enough bad weather in such circumstances.

5. The weather conditions associated with fast-moving cold fronts and squall lines are the most difficult to forecast accurately.
6. Errors in forecasting at the time of occurrence of bad weather are more prevalent than normal errors in forecasting whether bad weather will or will not occur within a span of time.
7. Surface visibility is more difficult to forecast than ceiling height, and snow reduces the visibility-forecasting problem to one of rather wild guesswork.

Forecasters can predict the following events at least 75 percent of the time: the passage of a fast-moving cold front or squall line within plus or minus 2 hours, as much as 10 hours in advance; the passage of warm fronts or slow-moving cold fronts within plus or minus 5 hours, up to 12 hours in advance; the rapid lowering of a ceiling below 1,000 feet in pre-warm front conditions to within plus or minus 200 feet, up to 4 hours in advance; the onset of a thunderstorm 1 or 2 hours in advance, if radar is available; the time rain or snow will begin within plus or minus 5 hours; the rapid deepening of a low-pressure center.

Some of the things that forecasters cannot predict with satisfactory results are: the time freezing rain will begin, the location and occurrence of extreme turbulence, the location and occurrence of heavy icing, the exact location and occurrence of a tornado, ceilings of 100 feet or less prior to their occurrence, the position of the center of a hurricane within 100 miles for 12 hours or more in advance, the time of occurrence of fog.

Weather systems and their parameters vary greatly in nature and rare events are more difficult to predict than are common events. Because of the rotation of the earth, weather conditions have pronounced periodic variations, as is demonstrated by the occurrence of fog during nighttime hours or the development of convective clouds and thunderstorms during the afternoon hours. It must be recalled that weather conditions often depend on the interaction of the wind, terrain, and surface conditions. Since many of the natural laws and variables are known, why can't perfect forecasts be made, at least short-term ones? The meteorologist knows only too well that many forecasts are badly bent or broken. But why, if we pretend to know and understand the laws of nature? If we can measure the various properties of the atmosphere, why don't perfect forecasts result? The answer to these questions is that we do not know values of *all* the meteorological parameters *everywhere* in the atmosphere. The observation network of weather stations from which the various meteorological data are collected are too widely dispersed to measure all of the variables. To define all the variations in time and space is nearly impossible. It is necessary to assume that each of the various meteorological parameters varies in time and space in a smooth regular fashion. In the real world, this does not happen because of the irregularity of the surface and its effect on the process of atmospheric heating. Energy exchange can be considered on a microscale, mesoscale, synoptic scale, and, finally, global scale. The microscale thermal and kinetic energy exchanges between the ocean and atmosphere and the earth and atmosphere are often ignored for economic reasons; approximations are often used instead. The errors introduced by these assumptions and approximations are magnified during the forecasting processes. It has been estimated that use of the dynamic kinematic techniques of forecasting results in an error magnification that will exceed the forecast signal in a 14-day period (which would be the limiting accuracy of the forecast). So the question must be raised, are better forecasts economically feasible? For how far into the future are forecasts worth their costs? These questions are societal, and answers are rare!

A pilot should have an understanding of the capabilities of the forecaster, but should never try to outguess one. A pilot should never try to influence the forecaster or briefer to describe the weather as better than it actually is believed to be. Finally, a pilot should never try to get the briefer to make a decision on whether flying is safe or not. This decision is the sole responsibility of the pilot!

The theme of this book can be stated simply: It is much better to let the atmosphere work for you than for you to work for the atmosphere. Even if you are an instrument-rated pilot, beware of situations that threaten to be borderline between VFR and IFR. These situations cause most of the in-flight weather-connected accidents. Maybe you have read the results of an accident investigation stating, "Continued VFR flight into IFR weather." Maybe not. Anyway, if you fly over the same area from trip to trip, roughly familiarize yourself with the geography, topography, and weather-reporting stations, and also the navigational aids in the area. Pilots must always know *where* they are!

An alert pilot will always be knowledgeable concerning the weather systems on or near his or her route, including the surface winds, ceilings, and visibilities at the landing fields along the route. When flying in mountainous areas, always be alert and aware of the density-altitude values expected at each of the air terminals.

Another point the alert pilot should be aware of is that the weather associated with a front or a feature of the weather chart probably won't look exactly like the illustrations of such conditions that you have seen in a book. No two weather situations are exactly alike. During any flight, monitor the weather broadcasts for the hourly weather reports and notice any *changes*. Be prepared for changes, especially in the forecast. Report what you observe. Familiarize yourself with average and extreme conditions by the season for the areas over which you fly.

Probably every pilot has heard this old saying: "There are old pilots and bold pilots." The implication, of course, is that pilots who are too bold never live to be too old! Learn to respect the weather! Be brave, but not foolish! And—happy flying!

BIBLIOGRAPHY

Boyne, W.J. *Flying: An Introduction to Flight, Airplanes, and Aviation Careers.* Englewood Cliffs, N.J.: Prentice-Hall, Inc., 1980.

Byers, H.R. *General Meteorology.* New York: McGraw-Hill, 1974.

Departments of Commerce, Defense, and Transportation. *Federal Meteorological Handbook*, No. 1. "Surface Observations," Chapter A6-3. Washington D.C.: U.S. Government Printing Office, 1979.

Departments of Transportation and Commerce. *Aviation Weather.* Washington D.C.: U.S. Government Printing Office, 1975.

Departments of Transportation and Commerce. *Aviation Weather Services.* Washington D.C.: U.S. Government Printing Office, 1975.

Dixon, P. *Soaring.* New York: Ballantine Books, 1970.

Flowers, W.D. *Sand Devils.* London: Meteorological Office Professional Notes, 1936.

Geiger, R. *The Climate Near the Ground.* Cambridge, Mass.: Harvard University Press, 1965.

Goody, R.M., and Walker, J.C.G. *Atmosphere.* Englewood Cliffs, N.J.: Prentice-Hall, Inc., 1972.

Huschke, R.E. *Glossary of Meteorology*. Boston: American Meteorological Society, 1959.

International Cloud Atlas. Geneva: World Meteorological Organization, 1956.

Ives, R.L. "Behavior of Dust Devils." *Bulletin of the American Meteorology Society* 28 (1947), 168–171.

Middleton, W.E.K. *Invention of Meteorological Instruments*. Baltimore, Md.: Johns Hopkins Press, 1969.

Perrie, D.W. *Cloud Physics*. New York: John Wiley & Sons, Inc., 1950.

Taylor, G.F. *Elementary Meteorology*. Englewood Cliffs, N.J.: Prentice-Hall, Inc., 1954.

U.S. Weather Bureau. *The Thunderstorm*. Washington, D.C.: U.S. Government Printing Office, 1949.

APPENDIX: GLOSSARY OF WEATHER TERMS

absolute instability the state of a *layer* within the atmosphere in which the vertical distribution of *temperature* is such that an *air parcel*, if given an upward or downward push, will move away from its initial level without further outside force being applied.

absolute temperature scale see *Kelvin temperature scale*.

absolute vorticity see *vorticity*.

adiabatic process the process by which fixed relationships are maintained during changes in *temperature*, volume, and pressure in a body of air without heat being added or removed from the body.

advection the horizontal transport of air or atmospheric properties. In *meteorology*, sometimes referred to as the horizontal component of *convection*.

advection fog fog resulting from the transport of warm, humid air over a cold surface.

The Glossary is reprinted from *Aviation Weather*, Departments of Transportation and Commerce (Washington, D.C.: U.S. Government Printing Office, 1975).

air density the mass density of the air in terms of weight per unit of volume.

air mass in *meteorology*, an extensive body of air within which the conditions of *temperature* and *moisture* in a horizontal plane are essentially uniform.

air-mass classification a system used to identify and to characterize different types of air masses according to a basic scheme. The system most commonly used classifies air masses primarily according to the thermal properties of their *source region:* tropical (T); polar (P); and Arctic or Antarctic (A). They are further classified according to *moisture* characteristics: continental (c) or maritime (m).

air parcel see *parcel.*

albedo the ratio of the amount of electromagnetic *radiation* reflected by a body to the amount incident upon it, commonly expressed in percentage. In *meteorology*, usually used in reference to *insolation (solar radiation).* For example, the albedo of wet sand is 9, meaning that about 9 percent of the incident insolation is reflected. Albedoes of other surfaces range upward to 80-85 for fresh *snow* cover; the average albedo for the earth and its atmosphere has been calculated to range between 35 and 43.

altimeter an instrument that determines the *altitude* of an object with respect to a fixed level; see *pressure altimeter.*

altimeter setting the value to which the scale of a *pressure altimeter* is set so as to read *true altitude* at field elevation.

altimeter-setting indicator a precision *aneroid barometer* calibrated to indicate directly the *altimeter setting.*

altitude height expressed in units of distance above a reference plane, usually above *mean sea level* or above ground.

1. *corrected altitude* the *indicated altitude* of an aircraft *altimeter* corrected for the *temperature* of the column of air below the aircraft, the correction being based on the estimated departure of existing temperature from standard atmospheric temperature; an approximation of *true altitude.*

2. *density altitude* the altitude in the *standard atmosphere* at which the air has the same *density* as the air at the point in question. An aircraft will have the same performance characteristics as it would have in a standard atmosphere at this altitude.

3. *indicated altitude* the altitude above mean sea level indicated on a *pressure altimeter* set at current local *altimeter setting*.

4. *pressure altitude* the altitude in the standard atmosphere at which the pressure is the same as at the point in question. Since an altimeter operates solely on pressure, this is the uncorrected altitude indicated by an altimeter set at standard sea level pressure of 29.92 inches of mercury or 1,013 millibars.

5. *radar altitude* the altitude of an aircraft determined by radar-type radio altimeter; thus, the actual distance from the nearest terrain or water feature encompassed by the downward-directed *radar beam*. For all practical purposes, it is the "actual" distance above a ground or inland water surface or the true altitude above an ocean surface.

6. *true altitude* the exact distance above mean sea level.

altocumulus white or gray layers or patches of cloud, often with a waved appearance. Cloud elements appear as rounded masses or rolls, composed mostly of liquid water droplets that may be supercooled; they may contain *ice crystals* at subfreezing temperatures.

altocumulus castellanus a species of middle cloud of which at least a fraction of the upper part presents some vertically developed, *cumuliform* protuberances (some of which are taller than they are wide, like castles, and which give the cloud a crenelated or turreted appearance — especially evident when seen from the side. Elements usually have a common base arranged in lines. This cloud indicates *instability* and *turbulence* at the altitudes of occurrence.

anemometer an instrument for measuring *wind speed*.

aneroid barometer a *barometer* that operates on the principle of having changing *atmospheric pressure* bend a metallic surface which, in turn, moves a pointer across a scale graduated in units of pressure.

angel in radar meteorology, an *echo* caused by physical phenomena not discernible to the eye. They have been observed when abnormally strong *gradients* of *temperature* and/or *moisture* are known to exist; sometimes attributed to insects or birds flying in the *radar beam*.

anomalous propagation sometimes called *AP*; in radar meteorology, the greater than normal bending of the *radar beam* such that an *echo* is received from a ground *target* at distances greater than normal *ground clutter.*

anticyclone an area of high *atmospheric pressure* which has a closed circulation that is anticyclonic, i.e., as viewed from above, the circulation is clockwise in the Northern Hemisphere, and undefined at the equator.

anvil cloud a popular name given to the top portion of a *cumulonimbus* cloud having an anvil-like form.

APOB a *sounding* made by an aircraft.

arctic air an *air mass* with characteristics developed mostly in winter over Arctic surfaces of ice and *snow.* Arctic air extends to great heights, and the surface temperatures are usually, but not always, colder than those of *polar air.*

arctic front the surface of discontinuity between very cold (Arctic) air flowing directly from the Arctic region and another less cold and, consequently, less dense *air mass.*

astronomical twilight see *twilight.*

atmosphere the mass of air surrounding the earth.

atmospheric pressure also called *barometric pressure;* the pressure exerted by the *atmosphere* as a consequence of gravitational attraction exerted upon the "column" of air lying directly above the point in question.

atmospherics disturbing effects produced in radio-receiving apparatus by atmospheric electrical phenomena, such as an electrical storm; static.

aurora a luminous emission over middle and high latitudes confined to the thin air of high altitudes and centered over the earth's magnetic poles. Called "aurora borealis" (northern lights) or "aurora australis" according to its occurrence in the Northern or Southern Hemisphere, respectively.

attenuation in radar meteorology, any process that reduces power density in radar signals.

1. *precipitation attenuation* reduction of power density because of absorption or reflection of energy by *precipitation.*

2. *range attenuation* reduction of radar power density because of distance from the antenna. It occurs in the outgoing signal at a rate proportional to $1/\text{range}^2$; the return signal is attenuated at the same rate.

backing shifting of the *wind* in a counterclockwise direction with respect to either space or time; the opposite of *veering*. Commonly used by meteorologists to refer to a cyclonic shift (counterclockwise in the Northern Hemisphere and clockwise in the Southern Hemisphere).

backscatter pertaining to *radar*, the energy reflected or scattered by a *target;* an *echo*.

banner cloud also called *cloud banner;* a bannerlike cloud streaming off from a mountain peak.

barograph a continuous-recording *barometer*.

barometer an instrument for measuring the pressure of the atmosphere; the two principal types are *mercurial* and *aneroid*.

barometric altimeter see *pressure altimeter*.

barometric pressure same as *atmospheric pressure*.

barometric tendency the change of *barometric pressure* within a specified period of time. In aviation-weather observations, routinely determined periodically, usually for a 3-hour period.

beam resolution see *resolution*.

Beaufort scale a classification scale of *wind speed*.

black blizzard same as *dust storm*.

blizzard a severe weather condition characterized by low temperatures and strong winds bearing a great amount of *snow*, either falling or picked up from the ground.

blowing dust a type of *lithometeor* composed of *dust* particles picked up locally from the surface by the wind and blown about in clouds and sheets.

blowing sand a type of *lithometeor* composed of sand picked up locally from the surface by the *wind* and carried to a height of 6 ft or more.

blowing snow a type of *hydrometeor* composed of *snow* picked up from the surface by the *wind* and carried to a height of 6 ft or more.

blowing spray a type of *hydrometeor* composed of water particles picked up by the *wind* from the surface of a large body of water.

bright band in radar meteorology, a narrow, intense *echo* on the *range-height indicator scope* resulting from water-covered ice particles of high reflectivity at the melting level.

Buys Ballot's law if an observer in the Northern Hemisphere faces away from the *wind* (back to the wind), the area of lower pressure is to the left.

calm the absence of *wind* or of apparent motion of the air.

cap cloud also called *cloud cap;* a standing or stationary caplike cloud crowning a mountain summit.

ceiling in *meteorology* in the United States, *(1)* the height above the surface of the base of the lowest *layer* of clouds or *obscuring phenomena* aloft that hides more than half of the sky; *(2)* the *vertical visibility* into an *obscuration;* see *summation principle.*

ceiling balloon a small balloon used to determine the height of a cloud base or the extent of *vertical visibility.*

ceiling light an instrument that projects a vertical light beam onto the base of a cloud or into surface-based *obscuring phenomena.* Used at night in conjunction with a *clinometer* to determine the height of the cloud base or as an aid in estimating *vertical visibility.*

ceilometer a cloud-height measuring system; it projects light on the cloud base, detects the reflection by a photoelectric cell, and determines height by triangulation.

Celsius (C) temperature scale a temperature scale with 0° as the melting point of pure ice and 100° as the boiling point of pure water at standard sea-level atmospheric pressure.

centigrade temperature scale same as *Celsius temperature scale.*

chaff pertaining to *radar, (1)* short, fine strips of metallic foil dropped from aircraft, usually by military forces, specifically for the purpose of jamming enemy radar; *(2)* applied loosely to the *echo* resulting from chaff.

change of state in *meteorology,* the transformation of water from one form to any other form. There are six possible transformations designated by the following five terms:

1. *condensation* the change of *water vapor* to liquid water.
2. *evaporation* the change of liquid water to water vapor.
3. *freezing* the change of liquid water to ice.
4. *melting* the change of ice to liquid water.
5. *sublimation* the change of (a) ice to water vapor or (b) water vapor to ice. See *latent heat.*

chinook a warm, dry *foehn* blowing down the eastern slopes of the Rocky Mountains over the adjacent plains in the United States and Canada.

cirriform all species and varieties of *cirrus, cirrocumulus,* and *cirrostratus* clouds; descriptive of clouds composed mostly or entirely of small *ice crystals,* usually transparent and white, often producing *halo* phenomena not observed with other cloud forms. The average height of such clouds ranges upward from 20,000 ft in middle latitudes.

cirrocumulus a *cirriform* cloud appearing as a thin sheet of small white puffs resembling flakes or patches of cotton without shadows; sometimes confused with *altocumulus.*

cirrostratus a *cirriform* cloud appearing as a whitish veil, usually fibrous, sometimes smooth; often produces *halo* phenomena; may totally cover the sky.

cirrus a *cirriform* cloud in the form of thin, white featherlike clouds in patches or narrow bands, with a fibrous and/or silky sheen; large *ice crystals* often trail downward a considerable vertical distance in fibrous, slanted, or irregularly curved wisps called "mares' tails."

civil twilight see *twilight.*

clear-air turbulence (CAT) *turbulence* encountered in air where no clouds are present; more popularly applied to high-level turbulence associated with *wind shear.*

clear icing or *clear ice* generally, the formation of a *layer* or mass of ice that is relatively transparent because of its homogeneous structure and the small number and size of air spaces. Synonymous with *glaze,* particularly with respect to aircraft icing; compare with *rime icing.* Factors that favor clear icing are large drop size, such as those found in *cumuliform* clouds, rapid accretion of *supercooled water,* and slow dissipation of *latent heat of fusion.*

climate the statistical collective of the weather conditions of a point or area during a specified interval of time (usually several decades); may be expressed in a variety of ways.

climatology the study of *climate.*

clinometer an instrument used in weather observation for measuring angles of inclination; it is used in conjunction with a *ceiling light* to determine cloud height at night.

cloud bank generally, a fairly well-defined mass of cloud observed at a distance; it covers an appreciable portion of the horizon sky, but does not extend overhead.

cloudburst in popular terminology, any sudden and heavy fall of *rain,* almost always of the *shower* type.

cloud cap see *cap cloud.*

cloud-detection radar a vertically directed radar to detect cloud bases and tops.

cold front any nonoccluded *front* that moves in such a way that colder air replaces warmer air.

condensation see *change of state.*

condensation level the height at which a rising *parcel* or *layer* of air would become saturated if lifted adiabatically.

condensation nuclei small particles in the air on which *water vapor* condenses or sublimates.

condensation trail or *contrail* also called *vapor trail;* a cloudlike streamer frequently observed to form behind aircraft flying in clear, cold, humid air.

conditionally unstable air unsaturated air that will become unstable when it becomes saturated; see *instability.*

conduction the transfer of heat by molecular action through a substance or from one substance in contact with another; transfer is always from warmer to colder temperatures.

constant-pressure chart a chart of a constant-pressure surface; it may contain analyses of height, *wind, temperature, humidity,* and/or other elements.

continental polar air see *polar air.*

continental tropical air see *tropical air.*

contour (1) In *meteorology,* a line of equal height on a *constant-pressure chart,* analogous to contours on a relief map. (2) In radar meteorology, a line on a radar scope of equal *echo* intensity.

contouring circuit on *weather radar,* a circuit that displays multiple contours of *echo* intensity simultaneously on the *plan-position indicator* or *range-height indicator scope;* see *contour (2).*

contrail contraction for *condensation trail.*

convection (1) In general, mass motions within a fluid resulting in transport and mixing of the properties of that fluid. (2) In *meteorology,* atmospheric motions that are predominantly vertical, resulting in vertical transport and mixing of atmospheric properties; distinguished from *advection.*

convective clouds see *cumuliform.*

convective condensation level (CCL) the lowest level at which *condensation* will occur as a result of *convection* due to surface heating. When condensation occurs at this level, the *layer* between the surface and the CCL will be thoroughly mixed, temperature *lapse rate* will be dry-adiabatic, and *mixing ratio* will be constant.

convective instability the state of an unsaturated *layer* of air whose *lapse rate* of temperature and moisture are such that when the air is lifted adiabatically until the layer becomes saturated, *convection* is spontaneous.

convergence the condition that exists when the pattern of *wind* within a given area is such that there is a net horizontal inflow of air into the area. In convergence at lower levels, the removal of the resulting excess is accomplished by an upward movement of air; consequently, areas of low-level convergent winds are regions favorable to the occurrence of clouds and *precipitation*. Compare with *divergence*.

coriolis force a deflective force resulting from the earth's rotation; it acts to the right of wind direction in the Northern Hemisphere and to the left in the Southern Hemisphere.

corona a prismatically colored circle (or arcs of a circle) with the sun or moon at its center. Coloration is from blue inside to red outside (opposite to that of a *halo*). It varies in size (much smaller), as opposed to the fixed diameter of the halo, and is characteristic of clouds composed of water droplets and valuable in differentiating between middle and *cirriform* clouds.

corposant see *St. Elmo's fire*.

corrected altitude see *altitude*.

cumuliform a term descriptive of all *convective clouds* exhibiting vertical development in constrast to the horizontally extended *stratiform* types.

cumulonimbus a *cumuliform* cloud type; it is heavy and dense, with considerable vertical extent in the form of massive towers, often with tops in the shape of an *anvil* or massive plume. Under the base of cumulonimbus, which often is very dark, there frequently exists *virga precipitation* and low, ragged clouds *(scud)*, either merged with it or not. Frequently accompanied by *lightning*, thunder, and sometimes *hail;* occasionally produces a *tornado* or a *waterspout*. The ultimate manifestation of the growth of a *cumulus* cloud, occasionally extending well into the *stratosphere*.

cumulonimbus mamma a *cumulonimbus* cloud with hanging protuberances, like pouches, festoons, or udders, on the under side of the cloud; usually indicative of severe *turbulence.*

cumulus a cloud in the form of individual detached domes or towers that are usually dense and well defined; develops vertically in the form of rising mounds, of which the bulging upper part often resembles a cauliflower; the sunlit parts of these clouds are mostly brilliant white; their bases are relatively dark and nearly horizontal.

cumulus fractus see *fractus.*

cyclogenesis any development or strengthening of cyclonic circulation in the *atmosphere.*

cyclone (1) An area of low *atmospheric pressure* which has a closed circulation that is cyclonic, i.e., as viewed from above, the circulation is counterclockwise in the Southern Hemisphere, undefined at the equator. Because cyclonic circulation and relatively low atmospheric pressure usually coexist, in common practice the terms cyclone and *low* are used interchangeably. Also, because cyclones often are accompanied by inclement (sometimes destructive) weather, they are frequently referred to simply as storms. (2) Frequently misused to denote a *tornado.* (3) In the Indian Ocean, a *tropical cyclone* of *hurricane* or *typhoon* force.

deepening a decrease in the central pressure of a pressure system. Usually applied to a *low* rather than to a *high*, although technically, it is acceptable in either sense; opposite of *filling.*

density (1) The ratio of the mass of any substance to the volume it occupies; weight per unit volume. (2) The ratio of any quantity to the volume or area it occupies, e.g., population per unit area, *power density*.

density-altitude see *altitude.*

depression in *meteorology*, an area of low pressure; a *low* or *trough*. Usually applied to a certain stage in the development of a *tropical cyclone*, to migratory lows and troughs, and to upper-level lows and troughs that are only weakly developed.

dew water condensed onto grass and other objects near the ground, the temperatures of which have fallen below the initial *dew point temperature* of the surface air, but are still above freezing; compare with *frost.*

dew point or *dew-point temperature;* the *temperature* to which a sample of air must be cooled, while the *mixing ratio* and *barometric pressure* remain constant, in order to attain *saturation* with respect to water.

discontinuity a zone with comparatively rapid transition of one or more meteorological elements.

disturbance in *meteorology*, applies rather loosely to *(1)* any low pressure or *cyclone*, but usually one that is relatively small in size; *(2)* an area where *weather, wind*, pressure, etc., show signs of cyclonic development; *(3)* any deviation in flow or pressure that is associated with a disturbed state of the weather, i.e., cloudiness and *precipitation;* and *(4)* any individual circulatory system within the primary circulation of the *atmosphere*.

diurnal daily, especially pertaining to a cycle completed within a 24-hour period, and which recurs every 24 hours.

divergence the condition that exists when the pattern of *wind* within a given area is such that there is a net horizontal flow of air outward from the region. In divergence at lower levels, the resulting deficiency is compensated for by *subsidence* of air from aloft; consequently, the air is heated and the relative *humidity* lowered, making divergence a warming and drying process. Low-level divergent regions are areas unfavorable to the occurrence of clouds and *precipitation*. The opposite of *convergence*.

doldrums the equatorial belt of calm or light and variable winds between the two trade-wind belts. Compare with *intertropical convergence zone*.

downdraft a relatively small-scale downward current of air; often observed on the lee side of large objects restricting the smooth flow of the air or in *precipitation* areas in or near *cumuliform* clouds.

drifting snow a type of *hydrometeor* composed of *snow* particles picked up from the surface, but carried to a height of less than 6 ft.

drizzle a form of *precipitation*; very small water drops that appear to float with the air currents while falling in an irregular path (unlike *rain*, which falls in a comparatively straight path, and unlike *fog* droplets, which remain suspended in the air).

dropsonde a *radiosonde* dropped by parachute from an aircraft to obtain *soundings* (measurements) of the *atmosphere* below.

dry-adiabatic lapse rate the rate of decrease of *temperature* with height when unsaturated air is lifted adiabatically (due to expansion as it is lifted to lower pressure). See *adiabatic process*.

dry-bulb a name given to an ordinary *thermometer* used to determine *temperature* of the air; also used as a contraction for *dry-bulb temperature*. Compare with *wet-bulb*.

dry-bulb temperature the temperature of the air.

dust a type of *lithometeor* composed of small earthen particles suspended in the *atmosphere*.

dust devil a small, vigorous *whirlwind*, usually of short duration, rendered visible by *dust*, sand, and debris picked up from the ground.

duster same as *dust storm*.

dust storm also called *duster, black blizzard*; an unusual, frequently severe weather condition characterized by strong *wind* and *dust*-filled air over an extensive area.

d-value departure of *true altitude* from *pressure altitude* (see *altitude*). Obtained by algebraically subtracting true altitude from pressure altitude; thus, it may be plus or minus. On a *constant-pressure chart*, the difference between actual height and standard atmospheric height of a constant-pressure surface.

echo in *radar*, (1) the energy reflected or scattered by a *target*. (2) the radar scope presentation of the return from a target.

eddy a local irregularity of *wind* in a larger scale wind flow. Small-scale eddies produce turbulent conditions.

estimated ceiling a ceiling classification applied when the ceiling height has been estimated by the observer or has been determined by some other method for which, because of the specified limits of time, distance, or *precipitation* conditions, a more descriptive classification cannot be applied.

evaporation see *change of state*.

extratropical low sometimes called extratropical cyclone, extratropical storm, any *cyclone* that is not a *tropical cyclone*, usually referring to the migratory frontal cyclones of middle and high latitudes.

eye the roughly circular area of calm or relatively light winds and comparatively fair weather at the center of a well-developed *tropical cyclone*. A *wall cloud* marks the outer boundary of the eye.

Fahrenheit (F) temperature scale a temperature scale with 32° as the melting point of pure ice and 212° as the boiling point of pure water at standard sea-level atmospheric pressure.

fall wind a cold *wind* blowing down-slope. Fall wind differs from *foehn* in that the air is initially cold enough to remain relatively cold despite compressional heating during descent.

filling an increase in the central pressure of a pressure system. More commonly applied to a *low* rather than a *high;* opposite of *deepening.*

first gust the leading edge of the spreading *downdraft, plow wind,* from an approaching *thunderstorm.*

flow line a *streamline.*

foehn a warm, dry down-slope wind; the warmness and dryness are due to adiabatic compression upon descent; characteristic of mountainous regions; see *adiabatic process, Chinook, Santa Ana.*

fog a *hydrometeor* consisting of numerous minute water droplets and based at the surface; droplets are small enough to be suspended in the earth's atmosphere indefinitely. Unlike *drizzle,* it does not fall to the surface. Fog differs from a cloud only in that a cloud is not based at the surface; it is distinguished from *haze* by its wetness and gray color.

fractus clouds in the form of irregular shreds, appearing torn; have a clearly ragged appearance; applies only to *stratus* and *cumulus* clouds, i.e., *cumulus fractus* and *stratus fractus.*

freezing see *change of state.*

freezing level a level in the atmosphere at which the temperature is 0°C (32°F).

front a surface, interface, or transition zone of discontinuity between two adjacent *air masses* of different densities; more simply, the boundary between two different air masses; see *frontal zone.*

frontal zone a *front* or zone with a marked increase of density gradient; used to denote that fronts are not truly a "surface" of discontinuity but rather a "zone" of rapid transition of meteorological elements.

frontogenesis the initial formation of a *front* or *frontal zone.*

frontolysis the dissipation of a *front.*

frost or *hoarfrost;* deposits of *ice crystals* formed by *sublimation* when the *temperature* and *dew point* are below freezing.

funnel cloud a *tornado* cloud or *vortex* cloud extending downward from the parent cloud but not reaching the ground.

glaze a coating of ice, generally clear and smooth, formed by freezing of supercooled water on a surface; see *clear icing*.

gradient in *meteorology*, a horizontal decrease in value per unit of distance of a parameter in the direction of maximum decrease; most commonly used with pressure, *temperature*, and *moisture*.

ground clutter pertaining to *radar*, a cluster of *echoes*, generally at short range, reflected from ground *targets*.

ground fog in the United States, a fog that conceals less than 0.6 of the sky and is not contiguous with the base of clouds.

gust a sudden brief increase in *wind*; according to the United States weather-observing practice, gusts are reported when the variation in *wind speed* between peaks and lulls is at least 10 knots.

hail a form of *precipitation* composed of balls or irregular lumps of ice, always produced by *convective clouds* that are nearly always *cumulonimbus*.

halo a prismatically colored or whitish circle (or arcs of a circle) with the sun or moon at its center. Coloration, if not white, is from red inside to blue outside (opposite to that of a *corona*). It is fixed in size, with an angular diameter of 22° (common) or 46° (rare), and is characteristic of clouds composed of *ice crystals*; valuable in differentiating between *cirriform* and lower clouds.

haze a type of *lithometeor* composed of fine *dust* or salt particles dispersed through a portion of the *atmosphere*; particles are so small they cannot be felt or individually seen with the naked eye (as compared with the larger particles of dust) but diminish the visibility; distinguished from *fog* by its bluish or yellowish tinge.

high also called a high-pressure system; an area of high *barometric pressure*, with its attendant system of *winds*; an *anticyclone*.

hoarfrost see *frost*.

humidity water vapor content of air; may be expressed as *specific humidity*, *relative humidity*, or *mixing ratio*.

hurricane a *tropical cyclone* in the Western Hemisphere with *winds* in excess of 65 knots or 120 kilometers per hour.

hydrometeor a general term for particles of liquid water or ice (such as *rain*, *fog*, *frost*), formed by modification of *water vapor* in the *atmosphere*; also water or ice particles lifted from the earth by the *wind* (such as sea spray or *blowing snow*).

hygrograph the record produced by a continuous-recording *hygrometer*.

hygrometer an instrument for measuring the *water vapor* content of the air.

ice crystals a type of *precipitation* composed of unbranched crystals in the form of needles, columns, or plates; usually having a very slight downward motion; may fall from a cloudless sky.

ice fog a type of fog composed of minute suspended particles of ice; occurs at a very low *temperature* and may cause *halo* phenomena.

ice needles a form of *ice crystals*.

ice pellets small, transparent or translucent, round or irregularly shaped pellets of ice; may be *(1)* hard grains that rebound on striking a hard surface, or *(2)* pellets of *snow* encased in ice.

icing in general, any deposit of ice forming on an object; see *clear icing, rime icing, glaze*.

indefinite ceiling a *ceiling* classification denoting *vertical visibility* into a surface based *obscuration*.

indicated altitude see *altitude*.

insolation incoming *solar radiation* falling upon the earth and the *atmosphere*.

instability a general term to indicate various states of the atmosphere in which spontaneous *convection* will occur when prescribed criteria are met; indicative of *turbulence;* see *absolute instability, conditionally unstable air, convective instability*.

intertropical convergence zone the boundary zone between the trade wind system of the Northern and Southern Hemispheres; it is characterized in maritime climates by showery *precipitations*, with *cumulonimbus* clouds sometimes extending to great heights.

inversion an increase in *temperature* with height; a reversal of the normal decrease with height in the *troposphere;* may also be applied to other meteorological properties.

isobar a line of equal or constant *barometric pressure*.

isoecho in *radar* circuitry, a circuit that reverses signal strength above a specified intensity level, thus causing a void on the scope in the most intense portion of an echo when maximum intensity is greater than the specified level.

isoheight on a *weather* chart, a line of equal height; same as *contour (1)*.

isoline a line of equal value of a variable quantity. For example, an isoline of *temperature* is an *isotherm;* see *isobar, isotach.*

isoshear a line of equal *wind shear.*

isotach a line of equal or constant *wind speed.*

isotherm a line of equal or constant *temperature.*

isothermal of equal or constant *temperature*, with respect to either space or time; more commonly, temperature with height; a zero *lapse rate.*

jet stream a quasi-horizontal stream of winds 50 knots or more concentrated within a narrow band embedded in the westerlies in the high *troposphere.*

katabatic wind any wind blowing down-slope; see *fall wind, foehn.*

Kelvin (K) temperature scale a temperature scale with 0° as the temperature at which all molecular motion ceases, i.e., absolute zero; $0°K = -273°C$. The Kelvin degree is identical to the Celsius degree; hence, at standard sea-level pressure, the melting point of pure ice is 273° and the boiling point of pure water is 373°.

knot a unit of speed equal to 1 nautical mile per hour.

land breeze a coastal breeze blowing from land to sea, caused by a *temperature* difference when the sea surface is warmer than the adjacent land; therefore, it usually blows at night and alternates with a *sea breeze*, which blows in the opposite direction by day.

lapse rate the rate of decrease of an atmospheric variable with height; commonly refers to decrease of *temperature* with height.

latent heat the amount of heat absorbed (converted to kinetic energy) during the processes of change of liquid water to *water vapor*, ice to water vapor, or ice to liquid water, or the reverse processes. Four basic classifications are:

1. *latent heat of condensation* heat released during the change of water vapor to water.

2. *latent heat of fusion* heat released during the change of water to ice, or the amount absorbed in the change of ice to water.
3. *latent heat of sublimation* heat released during the change of water vapor to ice, or the amount absorbed in the change of ice to water vapor.
4. *latent heat of vaporization* heat absorbed in the change of water to water vapor; the negative of latent heat of condensation.

layer in reference to sky cover, clouds or other *obscuring phenomena* whose bases are approximately at the same level; the layer may be continuous or composed of detached elements. The term does not imply that a clear space exists between layers or that the clouds or obscuring phenomena composing them are of the same type.

lee wave also called *mountain wave* or *standing wave;* any stationary wave disturbance caused by a barrier in a fluid flow. In the *atmosphere*, when sufficient *moisture* is present, this wave will be evidenced by *lenticular clouds* to the lee of mountain barriers.

lenticular cloud (or *lenticularis*) a species of cloud whose elements have the form of more or less isolated, generally smooth lenses or almonds. These clouds appear most often in formations of *orographic* origin, the result of *lee waves*, in which case they remain nearly stationary with respect to the terrain (*standing cloud*), but they also occur in regions without marked orography.

level of free convection (LFC) the level at which a *parcel* of air lifted dry-adiabatically until saturated and moist-adiabatically thereafter would become warmer than its surroundings in a conditionally unstable atmosphere; see *conditional instability* and *adiabatic process.*

lifting condensation level (LCL) the level at which a *parcel* of unsaturated air lifted dry-adiabatically would become saturated. Compare with *level of free convection* and *convective condensation level.*

lightning generally, any and all forms of visible electrical discharge produced by a *thunderstorm.*

lithometeor the general term for dry particles suspended in the *atmosphere*, such as *dust, haze, smoke,* and sand.

low also called a barometric *depression* or low-pressure system; an area of low *barometric pressure*, with its attendant system of *winds*; a *cyclone*.

mammato cumulus obsolete term for *cumulonimbus mamma*.

mares' tails see *cirrus*.

maritime polar air (mP) see *polar air*.

maritime tropical air (mT) see *tropical air*.

maximum wind axis on a constant-pressure chart, a line denoting the axis of maximum *wind speeds* at that constant-pressure surface.

mean sea level the average height of the surface of the sea for all stages of tide; used as a reference for elevations throughout the United States.

measured ceiling a *ceiling* classification applied when the ceiling value has been determined by instruments or the known heights of unobscured portions of objects, other than natural landmarks.

melting see *change of state*.

mercurial barometer a *barometer* in which pressure is determined by balancing air pressure against the weight of a column of mercury in an evacuated glass tube.

meteorological visibility in United States weather-observing practice, a main category of *visibility* which includes the subcategories of *prevailing visibility* and *runway visibility*. Meteorological visibility is a measure of horizontal visibility near the earth's surface, based on sighting of objects in the daytime or unfocused lights of moderate intensity at night. Compare with *slant visibility, runway visual range, vertical visibility*; see *surface visibility, tower visibility*, and *sector visibility*.

meteorology the science of the *atmosphere*.

microbarograph an aneroid *barograph* designed to record *atmospheric pressure* changes of very small magnitudes.

millibar (mb) an internationally used unit of pressure equal to 1,000 dynes per square centimeter; it is convenient for reporting *atmospheric pressure*.

mist a popular expression for *drizzle* or heavy *fog*.

mixing ratio the ratio by weight of the amount of *water vapor* in a volume of air to the amount of dry air; usually expressed as grams per kilogram (g/kg).

moist-adiabatic lapse rate see *saturated-adiabatic lapse rate*.

moisture an all-inclusive term denoting water in any or all of its three states.

monsoon a wind that in summer blows from sea to a continental interior, bringing copious *rain*, and in winter blows from the interior to the sea, resulting in sustained dry weather.

mountain wave a *standing wave* or *lee wave* to the lee of a mountain barrier.

nautical twilight see *twilight*.

negative vorticity see *vorticity*.

nimbostratus a principal cloud type, gray-colored, often dark, the appearance of which is rendered diffuse by more or less continuously falling *rain* or *snow*, which in most cases reaches the ground; it is thick enough throughout to blot out the sun.

noctilucent clouds clouds of unknown composition which occur at great heights, probably around 75 to 90 kilometers. They resemble thin *cirrus*, but usually have a bluish or silverish color, although sometimes orange to red, standing out against a dark night sky; rarely observed.

normal in *meteorology*, the value of an element averaged for a given location over a period of years and recognized as a standard.

numerical forecasting see *numerical weather prediction*.

numerical weather prediction forecasting by digital computers solving mathematical equations; used extensively in weather services throughout the world.

obscuration denotes sky hidden by surface-based *obscuring phenomena* and *vertical visibility* restricted overhead.

obscuring phenomena any *hydrometeor* or *lithometeor* other than clouds; may be surface-based or aloft.

occlusion same as *occluded front*.

occluded front commonly called an *occlusion*, or frontal occlusion; a composite of two fronts, as when a *cold front* overtakes a *warm front* or *quasi-stationary front*.

orographic of, pertaining to, or caused by mountains, as in orographic clouds, orographic lift, or orographic precipitation.

ozone an unstable form of oxygen; heaviest concentrations are in the *stratosphere;* corrosive to some metals; absorbs most ultraviolet *solar radiation*.

parcel a small volume of air, small enough to contain uniform distribution of its meteorological properties, and large enough to remain relatively self-contained and respond to all meteorological processes. (No specific dimensions have been defined, although the order of magnitude of 1 cubic foot has been suggested.)

partial obscuration a designation of sky cover when part of the sky is hidden by surface based *obscuring phenomena.*

pilot balloon a small free-lift balloon used to determine the speed and direction of *wind* in the upper air.

pilot-balloon observation commonly called *PIBAL*; a method of winds-aloft observation by visually tracking a *pilot balloon.*

plan-position indicator (PPI) scope a radar indicator scope displaying range and azimuth of *targets* in polar coordinates.

plow wind the spreading *downdraft* of a *thunderstorm;* a strong, straight-line *wind* in advance of the storm; see *first gust.*

polar air an *air mass* with characteristics developed over high latitudes, especially within the subpolar *highs.* Continental polar air (cP) has cold surface *temperatures,* low *moisture* content, and, especially in its source regions, great stability in the lower *layers.* It is shallow in comparison with *Arctic air.* Maritime polar air (mP) initially possesses similar properties to those of continental polar air, but in passing over warmer water it becomes unstable with a higher moisture content. Compare with *tropical air.*

polar front the semipermanent, semicontinuous front with separating *air masses* of tropical and polar origins.

positive vorticity see *vorticity.*

power density in radar meteorology, the amount of radiated energy per unit of cross-sectional area in the *radar beam.*

precipitation any or all forms of water particles, whether liquid or solid, that fall from the *atmosphere* and reach the surface. It is a major class of *hydrometeor,* distinguished from cloud and *virga* in that it must reach the surface.

precipitation attenuation see *attenuation.*

pressure see *atmospheric pressure.*

pressure altimeter an *aneroid barometer* with a scale graduated in *altitude* instead of pressure using standard *atmospheric pressure*-height relationships; shows indicated altitude (not necessarily true altitude); may be set to measure altitude (indicated) from any arbitrarily chosen level; see *altimeter setting, altitude.*

pressure altitude see *altitude.*

pressure gradient the rate of decrease of pressure per unit of distance at a fixed time.

pressure jump a sudden, significant increase in *station pressure.*

pressure tendency see *barometric tendency.*

prevailing easterlies the broad current or pattern of persistent easterly winds in the tropics and in polar regions.

prevailing visibility in the United States, the greatest horizontal visibility that is equaled or exceeded throughout the half of the horizon circle; it need not be a continuous half.

prevailing westerlies the dominant west-to-east motion of the *atmosphere* centered over middle latitudes of both hemispheres.

prevailing wind direction from which the wind blows most frequently.

prognostic chart usually contracted to *PROG;* a chart of expected or forecast conditions.

pseudo-adiabatic lapse rate see *saturated-adiabatic lapse rate.*

psychrometer an instrument consisting of a *wet-bulb* and a *dry-bulb thermometer* for measuring wet-bulb and dry-bulb temperature; used to determine *water vapor* content of the air.

pulse pertaining to *radar,* a brief burst of electromagnetic radiation emitted by the radar, of very short time duration; see *pulse length.*

pulse length pertaining to *radar,* the dimension of a radar pulse; may be expressed as the time duration or the length in linear units. Linear dimension is equal to time duration multiplied by the speed of propagation (approximately the speed of light).

quasi-stationary front commonly called *stationary front;* a front that is stationary or nearly so. Conventionally, a front moving at a speed of less than 5 knots is generally considered to be quasi-stationary.

radar contraction for radio detection and ranging; an electronic instrument used for the detection and ranging of distant objects of such composition that they scatter or reflect radio energy. Since *hydrometeors* can scatter radio energy, *weather radars*, operating on certain frequency bands, can detect the presence of *precipitation*, clouds, or both.

radar altitude see *altitude*.

radar beam the focused energy radiated by radar, similar to a flashlight or searchlight beam.

radar echo see *echo*.

radarsonde observation a *rawinsonde observation* in which *winds* are determined by *radar* tracking a balloon-borne *target*.

radiation the emission of energy by a medium and transferred either through free space or another medium, in the form of electromagnetic waves.

radiation fog fog characteristically resulting when radiational cooling of the earth's surface lowers the air *temperature* near the ground to or below its initial *dew point* on calm, clear nights.

radiosonde a balloon-borne instrument for measuring pressure, *temperature*, and *humidity* aloft.

radiosonde observation a *sounding* made by the instrument.

rain a form of *precipitation;* drops are larger than *drizzle* and fall in relatively straight, although not necessarily vertical, paths as compared to drizzle, which falls in irregular paths.

rain shower see *shower*.

range attenuation see *attenuation*.

range-height indicator (RHI) scope a *radar* indicator scope displaying a vertical cross-section of *targets* along a selected azimuth.

range resolution see *resolution*.

RAOB a *radiosonde observation*.

rawin a *rawinsonde observation*.

rawinsonde observation a combined winds-aloft and *radiosonde observation*. *Winds* are determined by tracking the *radiosonde* by radio direction-finder or *radar*.

refraction in *radar*, bending of the *radar beam* by variations in atmospheric density, *water vapor* content, and *temperature*.

1. *normal refraction* refraction of the radar beam under normal atmospheric conditions; normal radius of curvature of the beam is about four times the radius of curvature of the earth.

2. *super-refraction* more than normal bending of the radar beam resulting from abnormal vertical gradients of temperature and/or water vapor.

3. *subrefraction* less than normal bending of the radar beam resulting from abnormal vertical gradients of temperature and/or water vapor.

relative humidity the ratio of the existing amount of *water vapor* in the air at a given *temperature* to the maximum amount that could exist at that temperature; usually expressed in percent.

relative vorticity see *vorticity*.

remote scope in radar meteorology, a "slave" scope remoted from *weather radar*.

resolution pertaining to *radar*, the ability of radar to show discrete *targets* separately. The better the resolution, the closer two targets can be to each other and still be detected as separate targets.

1. *beam resolution* the ability of radar to distinguish between targets at approximately the same range but at different azimuths.

2. *range resolution* the ability of radar to distinguish between targets on the same azimuth but at different ranges.

ridge also called *ridge line;* in *meteorology*, an elongated area of relatively high *atmospheric pressure;* usually associated with and most clearly identified as an area of maximum anticyclonic curvature of the *wind* flow *(isobars, contours,* or *streamlines).*

rime icing or *rime ice;* the formation of a white or milky and opaque granular deposit of ice formed by the rapid freezing of supercooled water droplets as they impinge upon an exposed aircraft.

rocketsonde a type of *radiosonde* launched by a rocket and making its measurements during a parachute descent; capable of obtaining *soundings* to a much greater height than possible by balloon or aircraft.

roll cloud sometimes improperly called a *rotor cloud;* a dense and horizontal roll-shaped accessory cloud located on the lower leading edge of a *cumulonimbus* or, less often, a rapidly developing *cumulus;* indicative of *turbulence.*

rotor cloud sometimes improperly called a *roll cloud;* a turbulent cloud formation found in the lee of some large mountain *runway temperature* the temperature of the air just above a runway, ideally at engine and/or wing height, used in the determination of *density-altitude;* useful at airports when critical values of density-altitude prevail.

runway visibility the *meteorological visibility* along an identified runway determined from a specified point on the runway; may be determined by a *transmissometer* or by an observer.

runway visual range an instrumentally derived horizontal distance a pilot should see down the runway from the approach end; based on either the sighting of high-intensity runway lights or on the visual contrast of other objects, whichever yields the greatest visual range.

St. Elmo's fire also called *corposant;* a luminous brush discharge of electricity from protruding objects, such as masts and yardarms of ships, aircraft, lightning rods, steeples, etc., occurring in stormy weather.

Santa Ana a hot, dry, *foehn wind*, generally from the northeast or east, occurring west of the Sierra Nevada Mountains, especially in the pass and river valley near Santa Ana, California.

saturated-adiabatic lapse rate the rate of decrease of *temperature* with height as saturated air is lifted with no gain or loss of heat from outside sources; varies with temperature, being greatest at low temperatures; see *adiabatic process* and *dry-adiabatic lapse rate.*

saturation the condition of the *atmosphere* when actual *water vapor* present is the maximum possible at existing temperature.

scud small masses of *stratus fractus* clouds, usually *nimbostratus.*

sea breeze a coastal breeze blowing from sea to land, caused by the *temperature* difference when the land surface is warmer than the sea surface. Compare with *land breeze.*

sea fog a type of *advection fog* formed when air that has been lying over a warm surface is transported over a colder water surface.

sea-level pressure the *atmospheric pressure* at *mean sea level*, either directly measured by stations at sea level or empirically determined from the *station pressure* and *temperature* by stations not at sea level; used as a common reference for analyses of surface pressure patterns.

sea smoke same as *steam fog*.

sector visibility *meteorological visibility* within a specified sector of the horizon circle.

sensitivity time control a *radar* circuit designed to correct for range *attenuation* so that *echo* intensity on the scope is proportional to reflectivity of the *target* regardless of range.

shear see *wind shear*.

shower precipitation from a *cumuliform* cloud; characterized by the suddenness of beginning and ending, by the rapid change of intensity, and usually by rapid change in the appearance of the sky; showery precipitation may be in the form of *rain*, *ice pellets*, or *snow*.

slant visibility for an airborne observer, the distance at which he can see and distinguish objects on the ground.

sleet see *ice pellets*.

smog a mixture of *smoke* and *fog*.

smoke a restriction to *visibility* resulting from combustion.

snow precipitation composed of white or translucent *ice crystals*, chiefly in complex branched hexagonal form.

snow flurry popular term for *snow shower*, particularly of a very light and brief nature.

snow grains precipitation of very small, opaque grains of ice, similar in structure to *snow* crystals; the grains are fairly flat or elongated, with diameters generally less than 0.04 inch (1 mm).

snow pellets precipitation consisting of white, opaque, approximately round (sometimes conical) ice particles having a snow-like structure, and about 0.08 to 0.2 inch in diameter; crisp and easily crushed, differing in this respect from *snow grains*; rebound from a hard surface and often break up.

snow shower see *shower*.

solar radiation the total electromagnetic *radiation* emitted by the sun; see *insolation*.

sounding in *meteorology*, an upper-air observation; a *radiosonde observation*.

source region an extensive area of the earth's surface characterized by relatively uniform surface conditions where large masses of air remain long enough to take on characteristic *temperature* and *moisture* properties imparted by that surface.

specific humidity the ratio by weight of *water vapor* in a sample of air to the combined weight of water vapor and dry air. Compare with *mixing ratio*.

squall a sudden increase in *wind speed* by at least 15 knots to a peak of 20 knots or more and lasting for at least 1 minute. Essential difference between a *gust* and a squall is the duration of the peak speed.

squall line any nonfrontal line or narrow band of active *thunderstorms* (with or without *squalls*).

stability a state of the *atmosphere* in which the vertical distribution of *temperature* is such that a *parcel* will resist displacement from its initial level; see also *instability*.

standard atmosphere a hypothetical atmosphere based on climatological averages comprised of numerous physical constants, of which the most important are:

1. a surface *temperature* of 59°F (15°C) and a surface pressure of 29.92 inches of mercury (1013.2 millibars) at sea level.
2. a *lapse rate* in the *troposphere* of 6.5°C per kilometer (approximately 2°C per 1,000 ft).
3. a *tropopause* of 11 kilometers (approximately 36,000 ft) with a temperature of -56.5°C.
4. an *isothermal lapse rate* in the *stratosphere* to an *altitude* of 24 kilometers (approximately 80,000 ft).

standing cloud or standing lenticular altocumulus; see *lenticular cloud*.

standing wave a wave that remains stationary in a moving fluid. In aviation operations it is used most commonly to refer to a *lee wave* or *mountain wave*.

stationary front same as *quasi-stationary front*.

station pressure the actual *atmospheric pressure* at the observing station.

steam fog fog formed when cold air moves over relatively warm water or wet ground.

storm-detection radar a *weather radar* designed to detect *hydrometeors* of *precipitation* size; used primarily to detect storms with large drops or hailstones as opposed to clouds and light precipitation of small drop size.

stratiform descriptive term for clouds of extensive horizontal development, as contrasted to vertically developed *cumuliform* clouds; characteristic of stable air and, therefore, composed of small water droplets.

stratocumulus a low cloud, predominantly *stratiform*, in gray and/or whitish patches or *layers* that may or may not merge; elements are tessellated, rounded, or roll-shaped with relatively flat tops.

stratosphere the atmospheric *layer* above the *tropopause* (average altitude of base and top, 7 and 22 miles, respectively); characterized by a slight average increase of *temperature* from base to top and very stable; also characterized by low *moisture* content and absence of clouds.

stratus a low, gray cloud layer or sheet with fairly uniform base; sometimes appears in ragged patches; seldom produces *precipitation* but may produce *drizzle* or *snow grains*; a *stratiform* cloud.

stratus fractus see *fractus*.

streamline in *meteorology*, a line whose tangent is the *wind* direction at any point along the line; a *flow line*.

sublimation see *change of state*.

subrefraction see *refraction*.

subsidence a descending motion of air in the *atmosphere* over a rather broad area; usually associated with *divergence*.

summation principle the principle stating that the cover assigned to a *layer* is equal to the summation of the sky cover of the lowest layer plus the additional coverage at all successively higher layers up to and including the layer in question. Thus, no layer can be assigned a sky cover less than a lower layer, and no sky cover can be greater than 1.0 (10/10).

super-adiabatic lapse rate a *lapse rate* greater than the *dry-adiabatic lapse rate*; see *absolute instability*.

supercooled water liquid water at temperatures colder than freezing.

super-refraction see *refraction*.

surface inversion an inversion with its base at the surface, often caused by cooling of the air near the surface as a result of *terrestrial radiation*, especially at night.

surface visibility visibility observed from eye level above the ground.

synoptic chart a chart, such as the familiar weather map, which depicts the distribution of meteorological conditions over an area at a given time.

target in *radar*, any of the many types of objects detected by radar.

temperature in general, the degree of hotness or coldness as measured on some definite temperature scale by means of any of various types of *thermometers.*

temperature inversion see *inversion.*

terrestrial radiation the total infrared *radiation* emitted by the earth and the *atmosphere.*

thermograph a continuous-recording *thermometer.*

thermometer an instrument for measuring *temperature.*

theodolite in *meteorology,* an optical instrument that is used principally to observe the motion of a *pilot balloon.*

thunderstorm in general, a local storm invariably produced by a *cumulonimbus* cloud, and always accompanied by *lightning* and thunder.

tornado sometimes called *cyclone, twister;* a violently rotating column of air, pendant from a *cumulonimbus* cloud, and nearly always observable as "funnel-shaped." It is the most destructive of all small-scale atmospheric phenomena.

towering cumulus a rapidly growing *cumulus* in which height exceeds width.

tower visibility *prevailing visibility* determined from the control tower.

trade winds prevailing, almost continuous winds blowing with an easterly component from the subtropical high-pressure belts toward the *intertropical convergence zone;* northeast in the Northern Hemisphere, southeast in the Southern Hemisphere.

transmissometer an instrument system that shows the transmissivity of light through the *atmosphere.* Transmissivity may be translated either automatically or manually into *visibility* and/or *runway visual range.*

tropical air an *air mass* with characteristics developed over low latitudes. Maritime tropical air (mT), the principal type, is produced over the tropical and subtropical seas; very warm and humid. Continental tropical (cT) is produced over subtropical arid regions; hot and very dry. Compare with *polar air.*

tropical cyclone a general term for a *cyclone* that originates over tropical oceans. By international agreement, tropical cyclones have been classified according to their intensity, as follows:

1. *tropical depression* winds up to 34 knots (64 kilometers per hour).

2. *tropical storm* winds of 35 to 64 knots (65 to 119 kilometers per hour).
3. *hurricane* or *typhoon* winds of 65 knots or higher (120 kilometers per hour).

tropical depression see *tropical cyclone*.

tropical storm see *tropical cyclone*.

tropopause the transition zone between the *troposphere* and *stratosphere*, usually characterized by an abrupt change of *lapse rate*.

troposphere that portion of the *atmosphere* from the earth's surface to the *tropopause*, that is, the lowest 10 to 20 kilometers of the atmosphere. The troposphere is characterized by decreasing *temperature* with height, and by appreciable *water vapor*.

trough also called *trough line;* in *meteorology*, an elongated area of relatively low *atmospheric pressure;* usually associated with and most clearly identified as an area of maximum cyclonic curvature of the *wind* flow (*isobars, contours,* or *streamlines*). Compare with *ridge*.

true altitude see *altitude*.

true wind direction the direction, with respect to true north, from which the wind is blowing.

turbulence in *meteorology*, any irregular or disturbed flow in the *atmosphere*.

twilight the intervals of incomplete darkness following sunset and preceding sunrise. The time at which evening twilight ends or morning twilight begins is determined by arbitrary convention, and several kinds of twilight have been defined and used:

1. *civil twilight* the period of time before sunrise and after sunset when the sun is not more than 6° below the horizon.
2. *nautical twilight* the period of time before sunrise and after sunset when the sun is not more than 12° below the horizon.
3. *astronomical twilight* the period of time before sunrise and after sunset when the sun is not more than 18° below the horizon.

twister in the United States, a colloquial term for *tornado*.

typhoon a *tropical cyclone* in the Eastern Hemisphere with winds in excess of 65 knots (120 kilometers per hour).

undercast a cloud *layer* of 10/10 (1.0) coverage (to the nearest tenth), as viewed from an observation point above the layer.

unlimited ceiling a clear sky or a sky cover that does not meet the criteria for a *ceiling*.

unstable see *instability*.

updraft a localized upward current of air.

upper front a *front* aloft not extending to the earth's surface.

up-slope fog fog formed when air flows upward over rising terrain and is, consequently, adiabatically cooled to or below its initial *dew point*.

vapor pressure in *meteorology*, the pressure of *water vapor* in the atmosphere. Vapor pressure is that part of the total atmospheric pressure due to water vapor and is independent of the other atmospheric gases or vapors.

vapor trail same as *condensation trail*.

veering shifting of the *wind* in a clockwise direction with respect to either space or time; opposite of *backing*. Commonly used by meteorologists to refer to an anticyclonic shift (clockwise in the Northern Hemisphere and counterclockwise in the Southern Hemisphere).

vertical visibility the distance one can see upward into a surface based on *obscuration*, or the maximum height from which a pilot in flight can recognize the ground through a surface-based obscuration.

virga water or ice particles falling from a cloud, usually in wisps or streaks, and evaporating before reaching the ground.

visibility the greatest distance one can see and identify prominent objects.

visual range see *runway visual range*.

vortex in *meteorology*, any rotary flow in the *atmosphere*.

vorticity turning of the *atmosphere*. Vorticity may be imbedded in the total flow and not readily identified by a flow pattern.

1. *absolute vorticity* the rotation of the earth imparts vorticity to the atmosphere; absolute vorticity is the combined vorticity due to this rotation and vorticity due to circulation relative to the earth (relative vorticity).
2. *negative vorticity* vorticity caused by cyclonic turning; it is associated with downward motion of the air.
3. *positive vorticity* vorticity caused by cyclonic turning; it is associated with upward motion of the air.
4. *relative vorticity* vorticity of the air relative to the earth,

disregarding the component of vorticity resulting from earth's rotation.

wake turbulence turbulence found to the rear of a solid body in motion relative to a fluid; in aviation terminology, the turbulence caused by a moving aircraft.

wall cloud the well-defined bank of vertically developed clouds having a wall-like appearance which form the outer boundary of the *eye* of a well-developed *tropical cyclone.*

warm front any nonoccluded front which moves in such a way that warmer air replaces colder air.

warm sector the area covered by warm air at the surface and bounded by the *warm front* and *cold front* of a *wave cyclone.*

water equivalent the depth of water that would result from the melting of snow or ice.

water spout see *tornado.*

water vapor water in the invisible gaseous form.

wave cyclone a cyclone that forms and moves along a front. The circulation about the cyclone center tends to produce a wavelike deformation of the front.

weather the state of the *atmosphere*, mainly with respect to its effects on life and human activities; refers to instantaneous conditions or short-term changes, as opposed to *climate.*

weather radar radar specifically designed for observing weather; see *cloud-detection radar* and *storm-detection radar.*

weather vane   a *wind vane.*

wedge same as *ridge.*

wet-bulb contraction of either *wet-bulb temperature* or *wet-bulb thermometer.*

wet-bulb temperature the lowest temperature that can be obtained on a *wet-bulb thermometer* in any given sample of air, by evaporation of water (or ice) from the muslin wick; used in computing *dew point* and *relative humidity.*

wet-bulb thermometer a thermometer with a muslin-covered bulb used to measure *wet-bulb temperature.*

whirlwind a small, rotating column of air; may be visible as a *dust devil.*

willy-willy a *tropical cyclone* of hurricane strength near Australia.

wind air in motion relative to the surface of the earth; generally used to denote horizontal movement.

wind direction the direction from which wind is blowing.

wind shear the rate of change of *wind velocity* (direction and/or speed) per unit of distance; conventionally expressed as vertical or horizontal wind shear.

wind speed rate of wind movement in distance per unit of time.

wind vane an instrument to indicate wind direction.

wind velocity a vector term to include both *wind direction* and *wind speed.*

zonal wind a west wind; the westerly component of a wind; conventionally used to describe large-scale flow that is neither cyclonic nor anticyclonic.

Absolute atmospheric stability, 24-25
Absolute atmospheric instability, 23
Adiabatic expansion, atmospheric cooling and, 44
Adiabatic lapse rate, calculation of, 23
Adiabatic system, 21, 22
Advection, 18
Advection-radiation fog, 110
Aerometeorolograms, 10
Aircraft: altitude, temperature changes and, 33; stability, atmospheric stability and, 20-25; collisions, clouds and, 51; and condensation trails, 45; damage, turbulence and, 91; efficiency, icing and, 100-101; flight, 1, 6-7; meteorological sensors attached to wings of, 9-10. *See also* Aviation; Wind.
Airfields, at higher elevations, 115
Airfoil icing, 103, 104
Air-mass source region, 57-60
Air-mass thunderstorms. *See* Limited-state thunderstorms
Air masses, 57-61; classification of, 58-61; conditions for icing, 106; defined, 57; generation of, 57-58; life history of, 57; thermodynamic character of, 60; weather analysis and, 61-62. *See also* Fronts.
AIRMET advisories, 144
Airplane tow, vertical temperature sounding by, 133-34
Airports, fog and stratus cloud dispersal at, 46-46. *See also* Airfields.
Air pressure, oxygen absorption and, 6-7
Air Route Traffic Control Centers, 11, 139
Air Weather Service, 131
Altimeters: and atmospheric pressure, 30; and thunderstorm clouds, 89; standardization of, 30; units of pressure for, 28-29.
Altitude: of cumulonimbus clouds, 81-82; relationship to atmospheric pressure, 6.
Altocumulus clouds, 49, 50; frontal occlusions and, 72-73.
Altostratus clouds, 13, 49, 50; and flying hazards, 53, 54.
Aneroid barometer, 27-28
Anticyclone pressure systems, 35
Anticyclones: of primary air mass source regions, 58; weather associated with, 38-39.
Atmosphere: absolutely unstable, 76; composition of, 2; cooling of, 44; forecasting of thermals, 131-137; gases in, 2; general circulation of, 36-38; heating of, 2, 19; instability in troposphere, 5; layers, 3, 4; liquid in, 2; molecular weight calculation, 76; turbidity, temperature and, 20; vertical temperature structure, 3-7; weight of, 6-7. *See also* entries under *Atmospheric*.
Atmospheric density: determination of, 31; temperature changes and, 33-34.
Atmospheric heat, 14-25; and air-mass source region heat shift to, 57-58; atmospheric pressure and, 21; latitude and, 36; temperature lapse rate and, 20-25.
Atmospheric moisture, 43-47; buoyancy of air and, 76-77; clouds as index of, 69; condensation process, 44-45; and "meteors," 44; temperature and, 44; thermal convection and, 75-76; thunderstorms and, 78.
Atmospheric pollutants, smog and, 110-11
Atmospheric pressure, 26-42; altimeter settings and, 30; aneroid barometer and, 27-28; causes of variations in, 35, 36; discovery of, 26-27; discovery of barometer and, 27; law of hydrostatic equilibrium and, 31-32; relationship to altitude, 6-7; sea level differences and, 28-30, 34-35; temperature and, 21, 32-33; units used to denote, 28-29; uses of, 27; vertical pressure structure of, 29-30. *See also* Winds.
Atmospheric stability: clouds and, 25; criteria for, 23-25; temperature lapse rate and, 20-25.
Atmospheric temperatures: distribution of, 19-20; and volume relationships, atmospheric pressure and, 32-33.
Aviation: atmospheric moisture and, 43-56; atmospheric pressure and, 30; clouds and, 51, 69-71; and cold fronts, 71-72; and correction of altimeter settings, 33; and flying with the weather, 38; icing and, 104-5; latent heat and, 18-19; limited-state thunderstorms and, 83-84; local winds and, 42; occlusions and, 72-73; operations in U.S., 139-40; precipitation types and, 47; temperature effects on atmospheric density and, 33-34; thermodynamic character of air mass and, 60; and thunderstorms, 74, 86-87, 88-90; and vertical pressure structure of atmosphere, 29-30; weather knowledge and, 38-39, 144-45, 147-48. *See also* Clouds; Mountainous terrain flying.
Aviation weather assistance, 138-48; accidents and, 144-45; accuracy of forecasts and, 145-46; historical background of, 1, 138-39; minimum requirements, 141-42; National Weather Service and, 140; for nonstop trip, 144; pilot briefing, 140-42; U.S. Post Office and, 138-39; weather charts for, 142-44. *See also* Flight plans.

Balloons: and atmospheric pressure and temperature-volume relations, 32-33; description of thunderstorm during flight of, 78-81; and weather observations, 9; and wind measurements, 8, 9.
Barometer, 8, 27. *See also* Altimeter.
Bubble-type thermals, 126
Buoyancy: atmospheric stability and, 20-21; discovery of, 21-22; thermal convection and, 75-76; water vapor and, 76-77.

Carburetor, 103-4
CAT. *See* Clear-air turbulence
Ceilings, 46, 51; defined, 10; difference between vertical visibility and, 112; measurement of, 10.
Centrifugal force, weather and, 39
Cirrus clouds, 12, 50; cold fronts and, 71; debris, 74; warm fronts and, 68-69.
Civil Aeronautics Administration. *See* Federal Aviation Administration
Civil Air Patrol, 115
Clear-air turbulence (CAT), 93, 98
Clear ice, 101-102
Climactic zones, 19
Clouds: associated with fronts and occlusions, 66-67; atmospheric pressure and, 36; atmospheric temperature at location of, 52; cirriform, 49; cirrocumulus, 50; cirrostratus, 50; classification of, 11-13, 48-52; defined, 48; descriptions of, 50-51; flight safety and, 10-11, 13, 69-71; formation of, 45-47; in mountainous regions, 117; nimbus, 12-13; observation for weather forecasts, 11, 12, 47-56, 68-69; occlusions and, 72-73; as predictors of turbulence, 69; riding within, 69; and soaring, 125; stratocumulus, 13, 50-51.
Cold fronts, 60, 62; cyclones and, 64-65; flight and, 71-72; occlusions, 66-67; speed of, 63; thunderstorms and, 72.
Cold-core pressure systems, 35
Cold winds, 42
Colorado low, 67
Condensation, 44-45; thermal convection and, 75-77.
Condensation nuclei, 44-45; precipitation and, 46.
Condensation trails, 45
Conditional atmospheric stability or instability, 24
Conduction, 18
Continental air masses, 59

Continental-polar air-mass source region, 58
Convection, 18; soaring and, 127-28; thunderstorms and, 78, 86. *See also* Thermal convection.
Convective clouds, thermal soaring and, 130
Convective heat transfer, 57
Convective turbulence, 93
Coriolis effect, 38-39
Cumuliform clouds, 92; atmospheric instability and, 25; icing and, 105, 106; turbulence and, 52.
Cumulonimbus clouds, 13, 49, 51; cold fronts and, 71; flying hazards and, 55-56, 69-70; formation of, 77, 81-82. *See also* Thunderstorms.
Cumulus clouds, 12, 13, 51; thermals and, 130-31; turbulence and, 69.
Cyclone pressure systems, 35, 70
Cyclones, development along stationary fronts, 64-65

Deicing equipment, 102, 104
Density: Buoyancy principle and, 22-23; vertical structure of, 5-6.
Density altitude, 33-34; in mountainous regions, 115-16, 121.
Diurnal pressure variation, 35
Downdrafts: creation of, 82-83; head winds and, 119; of mature thunderstorms, 84-85; mountain waves and, 94-95.
Drizzle, 46; clouds predicting, 69; and IFR conditions, 111.
Dry-adiabatic lapse rate, 21, 23, 75, 76
Dust, 111, 113
Dust devils, 128-29
Dynamic high, 37

Earth: general circulation of atmosphere of, 36-38. *See also* Rotation of earth.
Electrometeors, 44
Energy, conversion of mass into, 14-15
Environmental lapse rate, dry-adiabatic lapse rate and, 23-24
Equatorial air mass, 59
Explosions, onboard, lightning and, 89

Federal Aviation Administration, 11, 139
Flight plans, 73; for cross-country flight, 118; and density-altitude problems, 116; icing and, 104; mountainous terrain and, 115, 120; weather conditions and, 56.
Flight Service Stations, 11, 140
Fog, 108-10; classifications, 108-10; dispersal, 46-47; formation of 46, 108-10, 113; in mountain flying, 117-18; size of particles, 46; and stratus clouds, 111.

Forecasting. *See* Aviation weather assistance
Freezing water particles, 45
Frictional force, 5, 39, 40
Frontal fogs, 110
Frontal lifts, soaring and, 135
Frontal occlusions, clouds and, 72-73
Frontal surface, 62-63
Frontal systems. *See* Fronts
Frontal waves, 64
Frontogenesis, 64, 68
Frontolyisis, 68
Fronts, 61-73; concept of, 61-62; emergence of, 62-63; life cycles of, 68; low-pressure systems and, 67-68; mountain flying and, 115, 122; rapid changes in, 73; types of, 63; weather sequence and, 68-69. *See also* Cold fronts; Polar fronts; Stationary fronts; Warm fronts.
Frost, 104
Fusion, energy of sun and, 14

General circulation theory, 39-42, 58
Glider flight. *See* Soaring
Gravity, atmospheric pressure and, 39
Ground heating, 5
Ground icing, 104

Hadley cell, 37
Hail: SIGMET advisories of, 143; size of hailstones, 47.
Hang gliding, 136-37
Haze, 111
Headwinds, flying into, 119
Heat: defined, 15; gases of atmosphere and, 2; measurement of, 15, 16; sources and sinks, 19-20. *See also* Atmospheric heat; Heat transfer; *and entries under* Thermal.
Heat transfer, 18-19; generation of air masses and, 57-58.
Heterogeneous nucleation, condensation and, 44-45
Heterosphere, 3, 4
High clouds, 12, 13, 49; flight safety and, 52-53; frontal cyclonic system and, 70.
High inversion fog, 110
High-pressure systems, low-pressure system conversion to, 68
Homogeneous nucleation, 45
Homosphere, 3, 4
Horizon line, in mountain flying, 118
Horizontal pressure-gradient force, weather and, 39
Hydrometeors, 44; thunderstorms and, 82.
Hygroscopic condensation nuclei, smog and, 110
Hydrostatic equation, 23
Hydrostatic equilibrium, law of, 31-32
Hypoxia, 7

Ice, coexistence with water, 46

Ice fog, 108
Icing, 100-107; categories of, 101-2; climbing out of, 105; in clouds, 52; conditions necessary for, 100-101, 106; effects of, 102-4; intensity classes, 102; and limited-state thunderstorms, 86-87; middle clouds and, 54; rate of, 103; safety precautions and, 104-6; terrain and, 106-7; thunderstorms and, 88. *See also* Structural icing.
IFR conditions, 107-13; avoidance of, 113; defined, 107; dust and haze, 111; fog, 108-10, 113; precipitation, 111-12; smog, 110-11; stratus clouds and, 111; VFR flight and, 107.
Inches of mercury, as pressure unit for altimeter settings, 28-29
Insolation, 5, 15; atmospheric temperature and, 19-20; radiation fog and, 109-10.
Instrument flight rules (IFR), visibility and, 10-11. *See also* VFR pilots.
Instrument icing, 103
International Civil Aviation Organization, 29
International Cloud Atlas, 13, 48, 49, 50-51
International Cloud Classification, 12-13
Isotherms, 19-20

Kernels. *See* Condensation nuclei
Kite meteography, 9, 139

Lakes, dust devils and, 129
Landings: density-altitude problems and, 116; iced up, 105; turbulence and, 91, 95-96.
Lapse rate, 5, 6
Latent heat, 18-19; condensation and, 45.
Latitude: air-mass classifications and, 59; atmospheric pressure and, 36; frequency of thunderstorms and, 85-86; heights of clouds, 11-12; origin of, 19.
Lift. *See* Soaring
Lightning, 74, 87-89
Limited-state thunderstorms, 79-81, 83-84, 86-87
Lithometeors, 44
Local winds, 40-42; soaring and, 127, 135-36.
Low clouds, 13, 49-50; aviation dangers and, 54-56.
Low-level significant weather prognosis charts, 143
Low-pressure systems, 67-68

Maritime air-mass source region, 58
Maritime air masses, 59
Mass: of atmosphere, 28; defined, 31; and heat content of matter, 16.

Matter, 15
Mechanical turbulence, 93-95; in mountainous areas, 116-17. See also Wake turbulence.
Meteorology: atmospheric pressure and, 28-30; development of, 7-8. See also Aviation weather assistance.
Meteors, 44
Middle clouds, 12, 13, 49; aviation safety and, 53, 54.
Migratory low-pressure systems, 38
Millibars, 28
Mixed ice, 101, 102
Moderate turbulence, 92
Moisture. See Atmospheric moisture
Molecules, 15; and heat content of matter, 16; temperature and, 15-16.
Monsoons, 40
Mountain passes, 119-20
Mountain-valley wind system, 40
Mountain waves, 94-95; orographic lifts and, 124-25.
Mountain weather, 120-22
Mountainous terrain, flying in, 114-22; atmospheric temperatures and, 19-20; clearance and, 114-15; clouds and, 117; conditions for, 121-22; corss-country soaring in, 134; density altitude, 115-16; and duration of limited-state thunderstorms, 85; flight plans and, 115; horizon line and, 118; and IFR conditions, 107-8; and interior valley fogs, 117-18; mountain winds and, 121-22; navigation equipment and, 120; nonfrontal low-pressure systems and, 67-68; sources of information on, 114; takeoffs in, 115; turbulence and, 94-95, 116-17; and valleys, 119; in VFR weather, 119-20; winds and, 118-19. See also Mountain weather.

National Meteorological Center (NMC), 11
National Weather Service, 140; historic background of, 138-39; and Stuve diagram, 131; and thermal forecasting, 133.
Natural objects, mechanical turbulence and, 93
Neutral atmospheric stability, 20, 24
Nevada low, 67
Nimbostratus clouds, 13, 49, 50; flying dangers, 54.
North Temperate Zone, 19

Occlusions: cloud systems and, 72-73; cold front, 66-67; warm front, 65-66.
Orographic clouds, in mountains, 120
Orographic lift, 124-25, 127
Oxygen: air pressure and, 6-7; as percentage of atmosphere, 2.

Photometeors, 44
Pilot briefing, 140-45
Pilot reports: on icing, 105-6; on turbulence, 92, 97-98, 99.
Planetary boundary layer of troposphere, 5
Polar air mass, 59
Polar easterlies, 37
Polar front, 38, 61-62, 138
Polar regions, high atmospheric pressure in, 37
Power plant icing, 18-19
Precipitation: altostratus clouds and, 53-54; below middle clouds, 53; condensation nuclei and, 46; icing and, 100; IFR conditions and, 111-12; nimbostratus clouds and, 54; thunderstorms and, 82-83; types of, 47.
Precipitation static, 89
Pressure-gradient force, 40
Pressure systems: changes in intensity of, 35; defined, 35-42; movement of, 35; vertical structure of, 5-6.
"Pseudodiabatic" chart, thermal index determination and, 131
Pseudo-cold front, 84-85

Radar summary charts, 142-43
Radiation, heat transfer and, 18
Radiation fog, 109-10
Radiosonde, 10, 139
Rain: clouds predicting, 69; visibility and, 111-12.
Rate-of-climb indicator, thunderstorms and, 89
Rawinsondes, 10
Relative humidity: condensation and, 45; thermal convection and, 75-76.
Rime ice, 101, 102
Roll clouds, 88
Rotation of earth: global pressure features and, 36; winds and, 36-37.
Runway length, atmospheric temperature and, 33-34. See also Takeoff.

Sailplanes. See Soaring
Saint Elmo's fire, 89
Salt solutions, formation of condensation kernel and, 45-46
Saturated adiabatic process, 21
Schaffer point, 46-47
Sea level, atmospheric pressure and, 28-30, 34-35
Sensible heat, 45
Service A weather reports, visibility in, 10-11
Showers: clouds predicting, 69; particles, 46.
SIGMET advisories, 143-44
Sirocco, 41
Skew T log P diagram, 131
Sleet, 47
Smog, 110-11
Snow: formation of, 45-46; IFR conditions and, 111.
Soaring, 123-37; auxilliary power sources and, 123; contribution to aviation, 136; forecasting of thermals, 131-37; mechanically initiated updrafts, 124-25; mountain waves and, 124-25; sources of power, 134-37; vertical motion of atmosphere and, 123-24. See also Thermal soaring.
Solar energy, 14-15; latitudinal differences in atmospheric pressure and, 36; thermals and, 126.
Solar radiation, calculation of intensity of, 14-15
Solids in atmosphere, 2
South Temperate Zone, 19
Special Theory of relativity, 14-15
Squall lines, tornadoes and, 87
Stalling speed, frost and, 104
Standard Atmosphere, density altitude and, 115, 116
Static system deicer, 105
Stationary fronts, 63-64, 65
Steady-state thunderstorm, 83, 84
Stratoform clouds: atmospheric stability and, 25; rime icing and, 102, 105.
Stratosphere, 4-5
Stratus clouds, 12, 13, 51, 111; dispersal of, 46-47; flying hazards, 55-56.
Structural icing: clouds and, 53, 54; effects of, 100-1, 103.
Stuve diagram, 131
Subpolar dynamic low, 37
Sun, energy source of, 14
Supersaturation, 45
Surface analysis, atmospheric pressure and, 34-35
Surface heating, thunderstorms and, 86

Takeoff: atmospheric temperature and, 33-34; density-altitude problems and, 116; at high elevations, 115; through water, 105; turbulence and, 91, 95-97.
Taxiing, through water, 105
Telegraph, meteorology and, 8
Temperature: air mass classification and, 59; of atmosphere at location of cloud, 52; atmospheric density and, 33-34; and atmospheric moisture retention, 44; defined, 15-16; for determining value of Thermal Index, 132-33; heat content of matter and, 16; icing and, 106; interior valley fogs and, 117-18; measurement of, 16-17; physiological reactions to, 16. See also Atmospheric temperature; Temperature inversions.
Temperature inversions: fog and, 108-9; smog and, 110-11.
Temperature lapse rate, atmospheric stability and, 20-25
Temperature scales, 17

Terrain: fog and, 110; icing and, 106–7; IFR conditions and, 107. *See also* Mountainous terrain.
Terrestrial long-wave radiation, troposphere and, 5
Thermal convection, thunderstorms and, 75–78
Thermal Index, determination of, 132–33
Thermal low, 68
Thermal soaring: clouds and, 130–31; cross-country in mountainous regions, 134; dust devils and, 128–29. *See also* Thermals.
"Thermal streets," 126–27
Thermal turbulence. *See* Convective turbulence
Thermal wind systems, 40–42
Thermals, 124–37; forecasting, 131–37; sizes and shapes of, 126–27.
Thermodynamic diagrams, 131
Thermodynamics: of air masses, 60; condensation and, 45; first law of, 22.
Thermometer, 8; development of, 17.
Thermosphere, 4
Thunder, lightning and, 74
Thunderstorms, 74–90; causes of, 75; clouds as signposts of, 51; cold fronts and, 72; cumulus stage, 77; definition of, 74–75; descriptions of, 78–81; dissipating stage of, 83; downdrafts, 84–85; and flying hazards, 88–90; frequency of, 85–86; hail and, 47; latent heat and, 45; mature stage, 83; requirements for, 85; SIGMET advisories and, 143; stages of, 77–78, 81–83; steady-state, 83, 84; thermal convection and, 75–78; turbulence and, 87–88; variations in, 74; weather radar and, 143. *See also* Limited state thunderstorms.
Topography: atmospheric pressure and, 36; atmospheric temperature and, 19.
Tornadoes: development of, 87; latent heat and, 45; SIGMET advisories and, 143; thunderstorms and, 64.
Torricellian tube, 27

Torrid Zone, 19
Trade winds, 36–38
Tropical air mass, 59
Tropical easterlies. *See* Trade winds
Tropical lows, 68
Troposphere, 3; condensation process in, 44–45; flight and, 4–5.
Turbulence, 91–99; active, 117; and atmospheric stability, 25; avoidance of, 99; classification of, 92–95; clouds and, 52, 117; cold fronts and, 71–72; cumulomammatus clouds and, 87; cumulonimbus clouds and, 55–56; extreme, 92; flying during, 70, 99; forecasting, 69, 92, 98; icing and, 88; intensity of, 91–92; latent heat and, 18; and lenticular clouds, 117; light, 92; and limited-state thunderstorms, 86; in mountainous terrains, 115, 116–17; passive, 116–17; and pendant cloud, 117; reporting criteria for, 92; severe, 92; SIGMET advisories of, 143–44; soaring and, 127–28; thunderstorms and, 79–81, 87–88; winds and, 43. *See also* Wake turbulence; Wind-shear turbulence.

United States Post Office, establishment of airways and, 138–39
United States Weather Bureau, 138. *See also* National Weather Service.
Updrafts: downdrafts and, 82–83; for soaring, 123–37.
Upslope fog, 110

Vacuums, 26
Vertical density structure, 5–6
Vertical pressure structure of atmosphere, 5–6, 29–30, 35–36
Vertical stability, defined, 20
Vertical temperature constant distribution (VTCD) of atmosphere, altimeter settings and, 30
Vertical temperature structure of atmosphere, 3–7

Vertical visibility, difference between ceiling and, 112
VFR pilots: flight plans and, 56, 73. *See also* IFR conditions.
Virga: below middle clouds, 53; thunderstorms and, 82.
Visibility, 10–11; clouds and, 51; thunderstorms and, 88. *See also* IFR conditions.
Visual flight rules. *See* VFR pilots

Wake turbulence, 93–94, 95–97
Warm air masses, 60
Warm-core high pressure system, 35
Warm-core low pressure system, 35
Warm fronts, 63, 64; cloud structure of, 70–71; cyclones and, 64–65; flying on top of, 70–71; flying under, 70; fog formation and, 113; occlusion and, 65–66; weather and, 68–69.
Wartime, secrecy of weather information during, 61
Water: coexistence with ice, 46; forms, 43; freezing and boiling points, 43. *See also* Atmospheric moisture.
Water vapor, atmospheric. *See* Atmospheric moisture
Wave-induced turbulence. *See* Clear-air turbulence
Weather charts, 63, 142
Weather forecasts, accuracy of, 145–47. *See also* Aviation weather assistance.
Weather instruments, development of, 8–10
Weather observations: first, 7–8; transmittal to pilot, 11.
Wind-shear turbulence, 88, 97–98
Wind towers, 8
Winds: Coriolis force and, 40; deflection, 5; dust and, 111; forces effecting, 39–40; influence of mountains on, 120–22; measurements of, 7–10; mountain flying and, 118–19; mountain waves and, 94–95; rotation of earth and, 36. *See also* Cold winds; Local winds; Trade winds.
World Meteorological Organization, 11, 13, 74